Simplified Reinforced Concrete

*Prentice-Hall International Series
in Civil Engineering and Engineering Mechanics*

William J. Hall, editor

Simplified Reinforced Concrete

Dr. Edward G. Nawy, P.E.

Distinguished Professor and Chairman
Department of Civil and Environmental Engineering
Rutgers University
The State University of New Jersey

PRENTICE-HALL, INC.
ENGLEWOOD CLIFFS, NEW JERSEY 07632

Library of Congress Cataloging-in-Publication Data

Nawy, Edward G.
    Simplified reinforced concrete.

    (Prentice-Hall international series in civil
engineering and engineering mechanics)
        Rev. ed. of: Reinforced concrete. c1985.
    Includes bibliographies and index.
        1. Reinforced concrete.   2. Reinforced concrete
construction.   I. Nawy, Edward G.   Reinforced
concrete.   II. Title.   III. Series.
    TA444.N38   1986   620.1'37   85–16884
    ISBN   0–13–810441–7

Editorial/production supervision and
  interior design: Nancy Milnamow
Cover design: Joseph Curcio
Manufacturing buyer: Rhett Conklin

Portions of this book were previously published under the title
Reinforced Concrete: A Fundamental Approach.

Printed in the United States of America

10   9   8   7   6   5   4   3   2   1

ISBN 0-13-810441-7   01

Prentice-Hall International, Inc., *London*
Prentice-Hall of Australia Pty. Limited, *Sydney*
Editora Prentice-Hall do Brasil, Ltda., *Rio de Janeiro*
Prentice-Hall Canada Inc., *Toronto*
Prentice-Hall of India Private Limited, *New Delhi*
Prentice-Hall of Japan, Inc., *Tokyo*
Prentice-Hall of Southeast Asia Pte. Ltd., *Singapore*
Whitehall Books Limited, *Wellington, New Zealand*

Contents

PREFACE   **xi**

Chapter 1
INTRODUCTION   **1**

    1.1   Historical Development of Structural Concrete   *1*
    1.2   Basic Hypothesis of Reinforced Concrete   *3*
    1.3   Analysis versus Design of Sections   *6*

Chapter 2
CONCRETE AND REINFORCEMENT   **11**

    2.1   Introduction   *11*
    2.2   Concrete-Producing Materials   *11*
    2.3   Criteria for Quality Control in Concrete   *17*
    2.4   Quality Tests on Concrete   *21*
    2.5   Placing and Curing of Concrete   *26*
    2.6   Properties of Hardened Concrete   *27*
    2.7   Shrinkage and Creep   *33*
    2.8   Reinforcement   *35*
    2.9   Concrete Structural Systems   *39*
          Selected References   *42*

Chapter 3
FLEXURAL ANALYSIS AND DESIGN   **43**

    3.1   Introduction   *43*
    3.2   ACI Load Factors and Safety Margins   *47*
    3.3   The Equivalent Rectangular Block   *49*
    3.4   Balanced Reinforcement Ratio $\bar{\rho}_b$ *55*
    3.5   Analysis of Singly Reinforced Rectangular Beams for Flexure   *56*
    3.6   One-Way Slabs   *67*
    3.7   Doubly Reinforced Beams   *70*
    3.8   Nonrectangular Sections (T and L Beams)   *80*
    3.9   Two-Way Slab and Plate Floor Systems   *94*
         Selected References   *96*
         Problems for Solution   *96*

Chapter 4
DEVELOPMENT LENGTH IN STEEL REINFORCEMENT   **100**

    4.1   Bond Stress Development   *100*
    4.2   Development of Flexural Reinforcement in Continuous Beams   *109*
    4.3   Splicing of Reinforcement   *111*
         Selected References   *116*
         Problems for Solution   *116*

Chapter 5
SHEAR AND TORSION   **118**

    5.1   Introduction   *118*
    5.2   Behavior of Homogeneous Beams   *118*
    5.3   Behavior of Reinforced Concrete Beams as Nonhomogeneous Sections   *123*
    5.4   Diagonal Tension Analysis of Slender and Intermediate Beams   *124*
    5.5   Web Steel Planar Truss Analogy   *126*
    5.6   Web Reinforcement Design Procedure for Shear   *130*
    5.7   Examples on the Design of Web Steel for Shear   *131*
    5.8   Torsion Theory   *136*
    5.9   Torsion in Reinforced Concrete Elements   *139*
    5.10 Design of Reinforced Concrete Beams Subjected to Combined Torsion,
          Bending, and Shear   *140*
         Selected References   *153*
         Problems for Solution   *154*

Chapter 6
DEFLECTION AND CRACKING   **157**

    6.1   Significance of Deflection Observation   *157*
    6.2   Deflection Behavior of Beams   *158*

6.3   Long-Term Deflection   *166*
6.4   Permissible Deflections in Beams and One-Way Slabs   *169*
6.5   Computation of Deflections   *171*
6.6   Deflection Control in One-Way Slabs   *178*
6.7   Flexural Cracking in Beams and One-Way Slabs   *181*
6.8   Permissible Crack Widths   *185*
6.9   Cracking Behavior and Crack Control in Two-Way-Action
       Slabs and Plates   *186*
       Selected References   *191*
       Problems for Solution   *192*

Chapter 7
COLUMNS   **195**

7.1   Introduction   *195*
7.2   Types of Columns   *196*
7.3   Strength of Short Concentrically Loaded Columns   *200*
7.4   Strength of Eccentrically Loaded Columns: Axial Load and
       Bending   *204*
7.5   Modes of Material Failure in Columns   *207*
7.6   Whitney's Approximate Solution   *219*
7.7   Column Strength Reduction Factor ϕ *220*
7.8   Load-Moment Strength Interaction Diagrams (P-M Diagrams)
       for Columns Controlled By Material Failure   *223*
7.9   Practical Design Considerations   *231*
7.10   Operational Procedure for the Design of Nonslender
       Columns   *233*
7.11   Numerical Examples for Analysis and Design of Nonslender
       Columns   *235*
       Selected References   *239*
       Problems for Solution   *240*

Chapter 8
FOOTINGS   **241**

8.1   Introduction   *241*
8.2   Types of Foundations   *243*
8.3   Shear and Flexural Behavior of Footings   *245*
8.4   Soil Bearing Pressure at Base of Footings   *246*
8.5   Design Considerations in Flexure   *251*
8.6   Design Considerations in Shear   *253*
8.7   Operational Procedure for the Design of Footings   *255*
8.8   Examples of Footing Design   *256*
       Selected References   *273*
       Problems for Solution   *274*

Appendix A
GEOMETRY

275

A.1   Selected Conversion Factors for SI Units   *276*
A.2   Geometrical Properties of Reinforcing Bars   *277*
A.3   Cross-Sectional Area of Bars for Various Bar
     Combinations   *278*
A.4   Area of Bars in a 1-Foot-Wide Slab Strip   *279*
A.5   Gross Moment of Inertia of T Sections   *280*

Appendix B
DETAILING OF REINFORCEMENT

281

B.1   Column Ties for Preassembled Lap-Spliced Cages   *282*
B.2   Column Ties for Standard Columns   *283*
B.3   Ties for Large and Special Columns   *284*
B.4   Corner and Joint Connection Details   *285*
B.5   Column Splice Details   *286*
B.6   Typical Working Drawing for a Parking Garage Structure   *287*
B.7   Typical Beam and Slab Reinforcing Working Drawing   *288*
B.8   Typical Working Drawing of Column Reinforcing Details   *289*
B.9   Elevator and Stairwell Details   *290*
B.10 Raft Foundation Details   *291*
B.11 Typical Reinforcement Bar Bending Schedule   *292*

Appendix C
CHARTS

293

C.1   Design Moment Strength ϕM_n for Slab Sections 12 in. Wide; $f'_c = 4000$ psi,
     $f_y = 40{,}000$ psi   *294*
C.2   Design Moment Strength ϕM_n for Beam Sections 10 in. Wide; $f'_c = 4000$ psi,
     $f_y = 40{,}000$ psi   *295*
C.3   Design Moment Strength ϕM_n for Slab Sections 12 in. Wide; $f'_c = 4000$ psi,
     $f_y = 60{,}000$ psi   *296*
C.4   Design Moment Strength ϕM_n for Beam Sections 10 in. Wide; $f'_c = 4000$ psi,
     $f_y = 60{,}000$ psi   *297*
C.5   Design Moment Strength ϕM_n for Slab Sections 12 in. Wide; $f'_c = 5000$ psi,
     $f_y = 60{,}000$ psi   *298*
C.6   Design Moment Strength ϕM_n for Beam Sections 10 in. Wide; $f'_c = 5000$ psi,
     $f_y = 60{,}000$ psi   *299*
C.7   Design Shear Strength ϕV_s of U Stirrups; $f_y = 40{,}000$ psi   *300*
C.8   Design Shear Strength ϕV_s of U Stirrups; $f_y = 60{,}000$ psi   *301*
C.9   Rectangular Columns Load–Moment Strength Interaction Diagrams;
     $f'_c = 4000$ psi, $f_y = 60{,}000$ psi, $\gamma = 0.75$ *302*
C.10 Rectangular Columns Load–Moment Strength Interaction Diagrams,
     $f'_c = 4000$ psi, $f_y = 60{,}000$ psi, $\gamma = 0.90$ *303*

C.11   Rectangular Columns Load–Moment Strength Interaction Diagrams,
$f'_c = 5000$ psi, $f_y = 60,000$ psi, $\gamma = 0.75$ *304*

C.12   Rectangular Columns Load–Moment Strength Interaction Diagrams,
$f'_c = 5000$ psi, $f_y = 60,000$ psi, $\gamma = 0.90$ *305*

C.13   Rectangular Columns Load–Moment Strength Interaction Diagrams,
$f'_c = 6000$ psi, $f_y = 60,000$ psi, $\gamma = 0.75$ *306*

C.14   Rectangular Columns Load–Moment Strength Interaction Diagrams,
$f'_c = 6000$ psi, $f_y = 60,000$ psi, $\gamma = 0.90$ *307*

C.15   Rectangular Columns Load–Moment Strength Interaction Diagrams,
$f'_c = 8000$ psi, $f_y = 60,000$ psi, $\gamma = 0.75$ *308*

C.16   Rectangular Columns Load–Moment Strength Interaction Diagrams,
$f'_c = 8000$ psi, $f_y = 60,000$ psi, $\gamma = 0.90$ *309*

INDEX

311

Preface

Reinforced concrete is a widely used material for constructed systems. Therefore graduates of every civil engineering program must, as a minimum requirement, have a basic knowledge of how to analyze and design concrete elements. Because design of elements of a total structure is achieved only by trial and adjustment, assuming a section and then analyzing it, design and analysis are combined in this book. Such an approach makes it simpler for the student first introduced to the subject of basic reinforced concrete design.

The text is the outgrowth of the author's lecture notes evolved while teaching the subject at Rutgers University over the past 25 years and the experience accumulated over the years in teaching and research in this area up to the Ph.D. level. The book is uniquely different from other textbooks at this level in that all its contents can be covered in one semester.

The concise discussion presented in Chapters 1 and 2 on concrete materials is intended to survey the principal features of how and why reinforced concrete is produced so that the reader can develop a feel for the constituent materials of a total structural system.

Since concrete is a nonelastic material, with the nonlinearity of its behavior starting at a very early stage of loading, only the ultimate-strength approach is presented in this book. Adequate coverage is given on the serviceability checks in terms of cracking and deflection behavior. In this manner the design should satisfy all the service-load-level requirements while ensuring that the theory used in the analysis (design) truly describes the actual behavior of the designed components.

Chapter 3 covers the flexural design and analysis of beams and one-way slabs with step-by-step trial and adjustment procedures and flowcharts in order to instill in the student the confidence necessary in following the most efficient route in design. The importance of this method of presenting the material throughout the text has become even more pronounced due to the increased use of personal computers in most structural design offices and the continuous emergence of computer programs in this area.

Chapter 4 discusses extensively the subject of development length and bond stress in the reinforcement. Chapter 5 covers the subject of shear, diagonal tension and torsion including a discussion of the fundamental behavior and the principal stress conditions in the structural element. Chapter 6 concisely addresses the subject of design for serviceability as required through deflection control and crack control in both beams and slabs. Chapter 7, on the analysis and design of columns and other compression members, extensively treats the subject of strain compatibility and strain distribution in a similar manner as in Chapter 3 (the flexural analysis and design of beams). Chapter 8 discusses the design of footings, thus completing the sequence of design steps. Conversions to SI units are included in the illustrative examples throughout the book.

An extensive appendix is included with several design handbook charts for detailing, as well as typical working drawings of a garage structure. The material in the appendix should serve as a general design aide for detailing the concrete elements and rapid choice of preliminary sections, thus assisting the instructor in introducing the student to such a design office approach.

Selected photographs of the various areas of structural behavior of concrete elements up to failure are included in all the chapters. They are taken from the published research work by the author with many of his M.S. and Ph.D. students at Rutgers University over the past two decades. Additionally, photographs of landmark structures throughout the United States are included to illustrate the versatility of design in reinforced concrete.

The textbook conforms to the provisions of ACI 318-83 with an eye to stressing the basics rather than tying every step to the code, which changes every six years. Consequently, no attempt was made to tie any design or analysis step to the particular equation number in the code but, rather, the student is expected to gain the habit of becoming familiar with the provisions and the section numbers of the ACI code on a separate basis, if that is necessary. In this manner the student can become well-versed with the ACI code as a dynamic, ever-changing document.

This book is primarily intended as a first course in basic design of reinforced concrete at state and junior colleges, architecture schools, and technical institutes. The presentation of an extensive number of flow charts on each of the subjects, as a particular feature of this book, should considerably simplify the instructor's task in presenting the topic, particularly if computer programming logic is also contemplated in this subject or in other subjects of the curriculum at the educational institution using the book.

Also, with the aid of the numerous photographs of elements tested to failure presented in the text, the instructor can explain the behavior modes of concrete members under the various stress types and conditions, thus giving the student a better feel for the basics of concrete design and performance. The contents should also serve as a valuable guideline to the designer who is interested in a concise treatment of the subject.

ACKNOWLEDGEMENT

Grateful acknowledgement is due to Dr. Edward J. Bloustein, President of Rutgers University, for his continuous support and encouragement over the years, to the American Concrete Institute for its contribution to the author's accomplishments and for permitting generous quotations of its ACI 318 Code and the illustrations from other ACI publications, and to his original mentor, Professor A. L. L. Baker of London University's Imperial College of Science and Technology who inspired him with the affection that he has developed for systems constructed of concrete.

Grateful acknowledgement is also made to the author's many students, both undergraduate and graduate, who had much to do in generating the writing of this book, to the many who assisted in his research activities over the years shown in the various photographs of laboratory tests throughout the book, and to Dr. P. N. Balaguru and engineers Mark J. Cipolloni, Robert M. Nawy, and Lily Sehayek for their combined valuable input to the original manuscript, and to engineer Regina S. Souza for her major input to the original manuscript and her extensive review of the present manuscript.

Particular thanks are due to engineer Carrie M. de Mackiewicz, one of the author's brightest former students, for her valuable and enthusiastic contribution to the present book which has made this author ever grateful.

Last but not least, grateful acknowledgement is due to Mr. Charles M. Iossi, Executive Editor and Vice President, Ms. Alice Dworkin, Executive Assistant, Ms. Nancy Milnamow, Production Editor, and Ms. A. Melissa Halverstadt, Marketing Manager, Prentice-Hall, for their continuous help and cooperation, and to the author's wife, Rachel E. Nawy, for her endurance through the process of developing this work.

Rutgers University **Edward G. Nawy**
The State University of New Jersey
New Brunswick, N.J. 08901

Simplified Reinforced Concrete

1

Introduction

1.1 HISTORICAL DEVELOPMENT OF STRUCTURAL CONCRETE

Concrete and its cementatious (volcanic) constituents, such as pozzolanic ash, have been used since the days of the Greeks, Romans, and possibly earlier ancient civilizations. The early part of the nineteenth century, however, marks the start of more intensive use of the material. In 1801 F. Coignet published his statement of principles of construction, recognizing the weakness of the material in tension. J. L. Lambot in 1850 constructed for the first time a small cement boat for exhibition in the 1855 World's Fair in Paris. J. Monier, a French gardener, patented metal frames as reinforcement for concrete garden plant containers in 1867 and Koenen in 1886 published the first manuscript on the theory and design of concrete structures. In 1906 C. A. P. Turner developed the first flat slab without beams.

Thereafter considerable progress occurred in this field so that by 1910 the German Committee for Reinforced Concrete, the Austrian Concrete Committee, the American Concrete Institute, and the British Concrete Institute were already established. Many buildings, bridges, and liquid containers of reinforced concrete were already constructed by 1920 and the era of linear and circular prestressing began.

The rapid developments in the art and science of reinforced and prestressed concrete analysis, design, and construction have resulted in very unique structural systems, such as Kresge Auditorium, Boston; 1951 Festival of Britain Dome; Marina Towers and Lake Point Tower, Chicago; Trump Towers, New York; and many, many others.

Ultimate-strength theories were codified in 1938 in the USSR and in 1956 in England and the United States. Limit theories also became a part of codes of several

Photo 1   Kennedy International Airport TWA Terminal, New York. (Courtesy of Ammann & Whitney.)

countries throughout the world. New constituent materials and composites of concrete are prevalent, including the high-strength concretes of a strength in compression up to 20,000 psi (137.9 MPa) and 1800 psi (12.41 MPa) in tension. Steel-reinforcing bars of strength in excess of 60,000 psi (413.7 MPa) and high-strength welded wire fabric in excess of 100,000 psi (689.5 MPa) ultimate strength are being used. Additionally, deformed bars of various forms have been produced. Such deformations help develop the maximum possible bond between the reinforcing bars and the surrounding concrete as a requisite for the viability of concrete as a structural medium. Prestressing steel of ultimate strengths in excess of 300,000 psi (2068 MPa) is available.

All these developments and the massive experimental and theoretical research that was conducted, particularly in the last two decades, resulted in rigorous theories and codes of practice. Consequently, a simplified approach is necessary to understanding the fundamental structural behavior of reinforced concrete elements.

Photo 2   The Trump Towers, Fifth Avenue, New York City. (Courtesy of Concrete Industry Board.)

1.2 BASIC HYPOTHESIS OF REINFORCED CONCRETE

Plain concrete is formed from a hardened mixture of cement, water, fine aggregate, coarse aggregate (crushed stone or gravel), air, and often other admixtures. The plastic mix is placed and consolidated in the formwork and then cured to facilitate the acceleration of the chemical hydration reaction of the water/cement mix, resulting in hardened concrete. The finished product has high compressive strength, and low resistance to tension, such that its tensile strength is approximately one-tenth of its

Photo 3   University of Illinois Assembly Hall at Urbana. (Courtesy of Ammann & Whitney.)

compressive strength. Consequently, tensile and shear reinforcement in the tensile regions of sections must be provided to compensate for the weak tension regions in the reinforced concrete element.

It is this deviation in the composition of a reinforced concrete section from the homogeneity of standard wood or steel sections that requires a modified approach to the basic principles of structural design, as will be explained in subsequent chapters. The two components of the heterogeneous reinforced concrete section are to be so arranged and proportioned that optimal use is made of the materials involved. Doing so is possible because concrete can easily be given any desired shape by placing and compacting the wet mixture of the constituent ingredients into suitable forms in which the plastic mass hardens. If the various ingredients are properly proportioned, the finished product becomes strong, durable, and in combination with the reinforcing bars, adaptable for use as main members of any structural system.

Photo 4   Afrikaans Languages Monument, Stellenbosch, South Africa (height of the main dynamically designed hollow columns, 186 ft).

Photo 5   Rockefeller Empire State Plaza, Albany, New York—Ammann & Whitney design. (Courtesy of New York Office of General Services.)

Photo 6 Empire State Performing Arts Center, Albany, New York—Ammann & Whitney design. (Courtesy of New York Office of General Services.)

1.3 ANALYSIS VERSUS DESIGN OF SECTIONS

From the foregoing discussion it is clear that a large number of parameters must be dealt with in proportioning a reinforced concrete element, such as geometrical width, depth, area of reinforcement, steel strain, concrete strain, and steel stress. Consequently, trial and adjustment is necessary in the choice of concrete sections, with assumptions based on conditions at site, availability of the constituent materials, particular demands of the owners, architectural and headroom requirements, the applicable codes, and environmental conditions. Such an array of parameters must be considered because of the fact that reinforced concrete is often a site-constructed composite, in contrast to the standard mill-fabricated beam and column sections in steel structures.

A trial section needs to be chosen for each critical location in a structural system. The trial section must be analyzed to determine if its nominal resisting strength is adequate to carry the applied factored load. Because more than one trial is often necessary to arrive at the required section, the first design input step generates into a series of trial-and-adjustment analyses.

The trial-and-adjustment procedures for the choice of a concrete section lead to the convergence of analysis and design. Thus every design is an analysis once a trial section is chosen. The availability of handbooks, charts, desktop and handheld personal computers and programs supports this approach as a more efficient, compact, and speedy instructional method compared with the traditional approach of treating the analysis of reinforced concrete separately from pure design.

Photo 7 Toronto City Hall, Toronto, Canada. (Courtesy of Portland Cement Association.)

Photo 8 Bridge ramp system. (Courtesy of Port Authority of New York and New Jersey.)

Photo 9   Sydney Opera House, Australia. (Courtesy of Australian Information Service.)

Photo 10   Priory Church, St. Louis, Missouri. (Courtesy of Portland Cement Association.)

Photo 11 Chicago Mercantile Exchange: a high strength concrete unique cantilever supports office tower. (Courtesy of Robert B. Johnson, Alfred Benesch and Co., Chicago.)

Photo 12 One Shell Plaza, New Orleans. (Courtesy of Portland Cement Association.)

2

Concrete and Reinforcement

2.1 INTRODUCTION

In order to obtain quality concrete for structural use, a knowledge of the concrete producing materials and their proportioning becomes essential. This chapter describes concrete-producing materials—namely, cement, fine and course aggregate, water, air, and admixtures. The cement manufacturing process, the composition of cement, the type and gradation of fine and course aggregate, and the function and importance of water and air are reviewed. In addition, the types and properties of reinforcement are described and a general overview of the basic concrete structural systems is given.

2.2 CONCRETE-PRODUCING MATERIALS

2.2.1 Portland Cement

Manufacture

Portland cement is made of finely powdered crystalline minerals composed primarily of calcium and aluminum silicates. Addition of water to these minerals produces a paste that, when hardened, becomes of stonelike strength. Its specific gravity ranges between 3.12 and 3.16 and it weighs 94 lb per cu ft, which is the unit weight of a commercial sack or bag of cement.

Photo 13 The Trump Towers under construction, Fifth Avenue, New York City.

The raw materials that form cement are

1. Lime (CaO)—from limestone
2. Silica (SiO_2)—from clay
3. Alumina (Al_2O_3)—from clay

(with very small percentages of Magnesia—namely, MgO—and sometimes some alkalis). Iron oxide is occasionally added to the mixture to aid in controlling its composition.

The manufacturing process (see Fig. 2.1) can be summarized as follows.

1. The raw mixture of CaO, SiO_2, and Al_2O_3 is ground with additional other minor ingredients either in dry or wet form. The wet form is called the "slurry" process.
2. The mixture is fed into the upper end of a slightly inclined rotary kiln.
3. As the heated kiln operates, the material passes from its upper to its lower end at a predetermined, controlled rate.

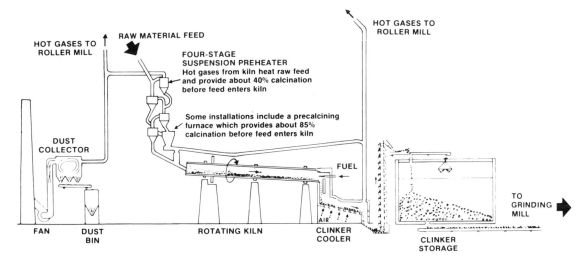

Figure 2.1   Portland cement manufacturing process. (From Ref. 2.13)

4. As the temperature of the mixture rises to the point of incipient fusion—that is, the *clinkering temperature*—it is kept at that temperature until the ingredients combine to form at 2700°F the portland cement pellet product. These pellets range in size from $\frac{1}{16}$ to 2 in. and are called clinkers.
5. The clinker is cooled and ground to a powdery form.
6. A small percentage of gypsum is added during the grinding process to control or retard the time of setting of cement in the field.
7. The final portland cement goes into silos for bulk shipment and, to a lesser extent, is packed in 94-lb bags for shipment.

Strength

The strength of cement is the result of a process of hydration. This process leads to a recrystallization in the form of interlocking crystals that produce the cement gel that has high compressive strength when it hardens. A study of Table 2.1 shows the relative contribution of each component of cement toward the rate of gain in strength. The early strength of portland cement is higher with higher percentages of C_3S. If moist curing is continuous, later strengths will be greater with higher percentages of C_2S. C_3A contributes to the strength developed during the first day after casting the concrete because it is the earliest to hydrate.

When portland cement combines with water during setting and hardening, lime is liberated from several compounds. The amount of lime liberated is approximately 20% by weight of the cement. Under unfavorable conditions this factor might cause disintegration of a structure owing to the leaching of lime from the cement. Such a condition should be prevented by the addition of a silicious mineral, such as pozzolan, to the cement.

TABLE 2.1 PERCENTAGE COMPOSITION OF PORTLAND CEMENTS

Type of cement	Component (%)							General characteristics
	C_3S	C_2S	C_3A	C_4AF	$CaSO_4$	CaO	MgO	
Normal: I	49	25	12	8	2.9	0.8	2.4	All-purpose cement
Modified: II	46	29	6	12	2.8	0.6	3.0	Comparative low heat liberation; used in large structures
High early strength: III	56	15	12	8	3.9	1.4	2.6	High strength in 3 days
Low heat: IV	30	46	5	13	2.9	0.3	2.7	Used in mass concrete dams
Sulfate resisting: V	43	36	4	12	2.7	0.4	1.6	Used in sewers and structures exposed to sulfates

The added mineral reacts with the lime in the presence of moisture to produce strong calcium silicate.

Influence of Cement on the Durability of Concrete

Disintegration of concrete due to cycles of wetting, freezing, thawing, and drying and the propagation of resulting cracks are a matter of great importance. The presence of minute air voids throughout the cement paste increases the resistance of concrete to disintegration. They can be achieved by the addition of air-entraining admixtures to the concrete while mixing.

Disintegration due to chemicals in contact with the structure, such as in port structures and substructures, can also be slowed down or prevented. Because the concrete in such cases is exposed to chlorides and sulphates of magnesium and sodium, it is imperative to specify sulphate-resisting cements. Usually type II cement will be adequate for use in seawater structures.

2.2.2 Water and Air

Water

Water is required in the production of concrete in order to precipitate chemical reaction with the cement, to wet the aggregate, and to lubricate the mixture for easy workability. Normally drinking water is used in mixing.

Water having harmful ingredients, contamination, silt, oil, sugar, or any other chemicals is destructive to the strength and setting properties of the cement paste and might adversely affect the workability of a mix.

Because colloidal gel or cement paste results merely from the chemical reaction between cement and water, it is not the proportion of water relative to the whole of the mixture of dry materials that is of concern in any study but only the proportion of water relative to the cement. Excessive water leaves an uneven honeycombed skeleton in the finished product after hydration has occurred whereas too little water prevents complete chemical reaction with the cement. The product, in both cases, is a concrete that is weaker and inferior to the one that is wanted.

Entrained Air

The gradual evaporation of excess water from the mix causes pores to be produced in the hardened concrete. If evenly distributed, these pores could give improved characteristics to the product. A very even distribution of pores by the artificial introduction of finely divided, uniformly distributed air bubbles throughout the product is possible by adding air-entraining agents, such as vinsol resin. Air entrainment increases workability and durability and reduces density, bleeding and segregation, and the required sand content in the mix. For these reasons, the percentage of entrained air should be kept at the required optimum value for the desired quality of the concrete. The optimum air content is 9% of the mortar fraction of the concrete. Air entraining in excess of 5 to 6% of the total mix starts to reduce the concrete strength proportionately.

Water/Cement Ratio

To summarize the preceding discussion, strict control should be maintained on the "water/cement ratio" and on the percentage of air in the mix. Because the water/cement ratio is considered the real measure of the strength of the concrete, it should be the criterion governing the design of most structural concretes. It is usually given as the ratio of weight of water to weight of cement in the mix.

2.2.3 Aggregates

Definition

Aggregates are those parts of the concrete that constitute the bulk of the finished product. They constitute 60 to 80% of the volume of the concrete and should be so graded that the whole mass of concrete acts as a relatively solid, homogeneous, dense combination, with the smaller sizes acting as an inert filler of the voids that exist between the larger particles.

Aggregates are of two types:

1. Coarse aggregate (gravel, crushed stone, or blast furnace slag)
2. Fine aggregate (natural or manufactured sand)

Because the aggregate constitutes the major part of the mix, the greater the amount of aggregate in the mix, the lower is the cost of the concrete, provided that the mix is of reasonable workability for the specific job in which it is used.

Coarse Aggregate

Coarse aggregate is classified as such if the smallest size of the particle is greater than $\frac{1}{4}$ in. (6 mm). Properties of the coarse aggregate affect the final strength of the hardened concrete and its resistance to disintegration, weathering, and other destructive effects. The mineral coarse aggregate must be clean from organic impurities and must have a good bond with the cement gel.

Here are the common types of coarse aggregates:

1. *Natural crushed stone*. It is produced by crushing natural stone or rock from quarries. The rock could be of igneous, sedimentary, or metamorphic type. Although crushed rock gives higher concrete strength, it is less workable in mixing and placing than other types.
2. *Natural gravel*. It is produced by the weathering action of running water on the beds and banks of streams. It gives less strength than crushed rock, but is more workable.
3. *Artificial coarse aggregates*. Such aggregates are mainly slag and expanded shale and are frequently used to produce lightweight concrete. They are the byproduct of other manufacturing processes, such as blast-furnace slag or expanded shale, or pumice for lightweight concrete.
4. *Heavyweight and nuclear-shielding aggregates*. As a result of the specific demands of our atomic age and the hazards of nuclear radiation due to the increasing number of atomic reactors and stations, it is necessary to produce special concretes as a shield against x rays, gamma rays, and neutrons. Economic and workability factors are not of prime importance in such concretes. The main heavy coarse aggregate types are: steel punchings, barites, magnatites, and limonites.

Concrete with ordinary aggregate weighs about 144 lb per cu ft, but concrete made with these heavy aggregates weighs from 225 to 330 lb per cu ft. The property of heavyweight radiation-shielding concrete depends on the density of the compact product rather than primarily on the water/cement ratio criterion. In certain cases, high density is the only consideration, whereas both density and strength are important in others.

Fine Aggregate

Fine aggregate is the smaller-sized filler, consisting generally of sand. It ranges in size from no. 4 to no. 100 U.S. standard sieves. A good fine aggregate should always be clean from organic impurities, clay, or any deleterious material or excessive filler of size smaller than no. 100 sieve. Preferably it should have a well-graded combination conforming to the ASTM sieve analysis standards. For radiation-shielding concrete, fine steel shot as well as crushed iron ore are used as fine aggregate.

The recommended gradings of the coarse and fine aggregates are given in detail in ASTM standards C330 and C637 and shown in Tables 2.2 to 2.5.

TABLE 2.2 GRADING REQUIREMENTS FOR AGGREGATES IN NORMALWEIGHT CONCRETE

| U.S. standard sieve size | Percent passing | | | | |
| | Coarse aggregate | | | | Fine aggregate |
	No. 4 to 2 in.	No. 4 to $1\frac{1}{2}$ in.	No. 4 to 1 in.	No. 4 to $\frac{3}{4}$ in.	
2 in.	95–100	100	—	—	—
$1\frac{1}{2}$ in.	—	95–100	100	—	—
1 in.	25–70	—	95–100	100	—
$\frac{3}{4}$ in.	—	35–70	—	90–100	—
$\frac{1}{2}$ in.	10–30	—	25–60	—	—
$\frac{3}{8}$ in.	—	10–30	—	20–55	100
No. 4	0–5	0–5	0–10	0–10	95–100
No. 8	0	0	0–5	0–5	80–100
No. 16	0	0	0	0	50–85
No. 30	0	0	0	0	25–60
No. 50	0	0	0	0	10–30
No. 100	0	0	0	0	2–10

2.2.4 Admixtures

Admixtures are materials other than water, aggregate, or hydraulic cement that are used as ingredients of concrete and that are added to the batch immediately before or during the mixing. Their function is to modify the properties of the concrete so as "to make it more suitable for the work at hand, or for economy, or for other purposes such as saving energy" (Ref. 2.6). The major types of admixtures can be summarized as follows:

1. Accelerating admixtures
2. Air-entraining admixtures
3. Water-reducing admixtures and set-controlling admixtures
4. Finely divided mineral admixtures
5. Admixtures for no-slump concretes
6. Polymers
7. Superplasticizers

2.3 CRITERIA FOR QUALITY CONCRETE

The general characteristics of quality concrete are described next.

TABLE 2.3 GRADING REQUIREMENTS FOR AGGREGATES IN LIGHTWEIGHT STRUCTURAL CONCRETE

Size designation	Percentages (by weight) passing sieves having square openings								
	1 in. (25.0 mm)	$\frac{3}{4}$ in. (19.0 mm)	$\frac{1}{2}$ in. (12.5 mm)	$\frac{3}{8}$ in. (9.5 mm)	No. 4 (4.75 mm)	No. 8 (2.36 mm)	No. 16 (1.18 mm)	No. 50 (300 μm)	No. 100 (150 μm)
Fine aggregate No. 4 to 0	—	—	—	100	85–100	—	40–80	10–35	5–25
Coarse aggregate									
1 in. to No. 4	95–100	90–100	25–60	—	0–10	—	—	—	—
$\frac{3}{4}$ in. to No. 4	100	90–100	—	10–50	0–15	—	—	—	—
$\frac{1}{2}$ in. to No. 4	—	100	90–100	40–80	0–20	0–10	—	—	—
$\frac{3}{8}$ in. to No. 8	—	—	100	80–100	5–40	0–20	0–10	—	—
Combined fine and coarse aggregate									
$\frac{1}{2}$ in. to 0	—	100	95–100	—	50–80	—	—	5–20	2–15
$\frac{3}{8}$ in. to 0	—	—	100	90–100	65–90	35–65	—	10–25	5–15

TABLE 2.4   GRADING REQUIREMENTS FOR COARSE AGGREGATE
FOR AGGREGATE CONCRETE

	Percentage passing	
Sieve size	Grading 1: for $1\frac{1}{2}$ in. (37.5 mm) maximum-size aggregate	Grading 2: for $\frac{3}{4}$ in. (19.0 mm) maximum-size aggregate
Coarse Aggregate		
2 in. (50 mm)	100	—
$1\frac{1}{2}$ in. (37.5 mm)	95–100	100
1 in. (25.0 mm)	40–80	95–100
$\frac{3}{4}$ in. (19.0 mm)	20–45	40–80
$\frac{1}{2}$ in (12.5 mm)	0–10	0–15
$\frac{3}{8}$ in. (9.5 mm)	0–2	0–2
Fine Aggregate		
No. 8 (2.36 mm)	100	—
No. 16 (1.18 mm)	95–100	75–95
No. 30 (600 μm)	55–80	75–95
No. 50 (300 μm)	30–55	45–65
No. 100 (150 μm)	10–30	20–40
No. 200 (75 μm)	0–10	0–10
Fineness modulus	1.30–2.10	1.00–1.60

Data in Tables 2.2 to 2.4 reprinted with permission from the American Society
for Testing and Materials, Phildadelphia, PA.

TABLE 2.5   UNIT WEIGHT OF AGGREGATES

Type	Unit weight of dry-rodded aggregate (lb/ft^3)^a	Unit weight of concrete (lb/ft^3)^a
Insulating concretes (perlite, vermiculite, etc.)	15–50	20–90
Structural lightweight	40–70	90–110
Normalweight	70–110	130–160
Heavyweight	>135	180–380

^{a} 1 lb/ft^3 = 16.02 kg/m^3.

Compactness

The space occupied by the concrete should, as much as possible, be filled with solid
aggregate and cement gel free from honeycombing. Compactness may be the primary
criterion for those types of concrete that intercept nuclear radiation.

Strength

Concrete should always have sufficient strength and internal resistance to the various types of failure that it can suffer.

Water/Cement Ratio

This ratio should be suitably controlled to give the required compressive and tensile design strengths.

Texture

Exposed concrete surfaces should have dense and hard textures that can withstand adverse weather conditions.

In order to achieve these properties, quality control and quality assurance should be rigorously maintained in selecting and processing the following parameters:

1. Quality of cement
2. Proportion of cement in relation to water in the mix
3. Strength and cleanliness of aggregate
4. Interaction or adhesion between cement paste and aggregate
5. Adequate mixing of the ingredients
6. Proper placing, finishing, and compaction of the fresh concrete
7. Curing at temperature not below 50°F while the placed concrete gains strength

A glance at these requirements shows that most control actions should be taken prior to placing the fresh concrete. Because such controls are governed by the proportions and the mechanical ease or difficulty in handling and placing, the development of criteria based on the theory of proportioning for each mix should be considered.

The most accepted process of proportioning concrete mixes is the American Concrete Institute's method for both normalweight and lightweight concretes. Besides designing a mix to achieve the prescribed level of 28-day compressive strength, the purpose of mix design is also to produce workable concrete that is easy to place in the forms. A measure of the degree of consistency and extent of workability is the slump. In the slump test, the plastic concrete specimen is formed into a conical metal mold as described in ASTM standard C143. The mold is lifted, leaving the concrete to "slump"—that is, spread or drop in height. This drop in height is the slump measure of the degree of workability of the mix.

Mix Designs for Nuclear-Shielding Concrete

From the foregoing discussion, the design criterion for normal concrete is the water/cement ratio. In concrete used for shielding against x rays, gamma rays, and neutrons, however, the criterion used is compactness or density of mix regardless of workability. In order to achieve maximum density, tests have been conducted on various mixes by using crushed magnatite ore or fine steel shot instead of sand and steel punchings, magnatites, barites, or limonites instead of stone as discussed previously.

Results of these tests for both compactness and strength have shown that the water/cement ratio should be limited to 3.5 to 4.5 gallons of water per cubic yard.

Proportioning Theory

In order to achieve the quality and strength required for a particular mix, the ACI proportioning theory, based on the water/cement approach, is recommended.

The water/cement ratio (W/C ratio) theory states that for a given combination of materials and as long as workable consistency is obtained, the strength of concrete at a given age depends on the ratio of the weight of mixing water to the weight of cement. In other words, if the ratio of water to cement is fixed, the strength of concrete at a certain age is also essentially fixed as long as the mixture is plastic and workable and the aggregate sound, durable, and free of deleterious materials. Although strength depends on the W/C ratio, economy depends on the percentage of aggregate present that would still give a workable mix. The aim of the designer should always be to get concrete mixtures of optimum strength at minimum cement content and acceptable workability. The lower the W/C ratio, the higher is the concrete strength.

Once the W/C ratio is established and the workability or consistency needed for the specific design is chosen, the rest of the mix design process should be simple manipulation with diagrams and tables based on large numbers of trial mixes. Such diagrams and tables allow an estimate of the required mix proportions for various conditions and permit predetermination based on tests of small, unrepresentative batches.

In order to apply this mix proportioning theory, proportions of the mix ingredients are selected from the appropriate ACI recommended proportions given in Tables 2.6, 2.7, and 2.8.

2.4 QUALITY TESTS ON CONCRETE

Workability or Consistency

1. Slump test by means of the standard ASTM code. The slump in inches recorded for the mix indicates its workability. (See photo 15.)
2. Remolding tests using Power's flow table.
3. Kelley's ball apparatus

Air Content

Measurement of air content on fresh concrete is always necessary, especially when air-entraining agents are used.

Compressive Strength of Hardened Concrete

This is done by loading cylinders 6 in. in diameter and 12 in. in height perpendicular to the axis of the cylinder.

TABLE 2.6 APPROXIMATE MIXING WATER AND AIR CONTENT REQUIREMENTS FOR DIFFERENT SLUMPS AND NOMINAL MAXIMUM SIZES OF AGGREGATES

Slump (in.)	Water (lb/yd³ of concrete for indicated nominal maximum sizes of aggregate)							
	$\frac{3}{8}$ in.[a]	$\frac{1}{2}$ in.[a]	$\frac{3}{4}$ in.[a]	1 in.[a]	$1\frac{1}{2}$ in.[a]	2 in.[a,b]	3 in.[b,c]	6 in.[b,c]
Non-Air-Entrained Concrete								
1 to 2	350	335	315	300	275	260	220	190
3 to 4	385	365	340	325	300	285	245	210
6 to 7	410	385	360	340	315	300	270	—
Approximate amount of entrapped air in non-air-entrained concrete (%)	3	2.5	2	1.5	1	0.5	0.3	0.2
Air-Entrained Concrete								
1 to 2	305	295	280	270	250	240	205	180
3 to 4	340	325	305	295	275	265	225	200
6 to 7	365	345	325	310	290	280	260	—
Recommended average total air content[d] (percent for level of exposure)								
Mild exposure	4.5	4.0	3.5	3.0	2.5	2.0	1.5[e,f]	1.0[e,f]
Moderate exposure	6.0	5.5	5.0	4.5	4.5	4.0	3.5[e,f]	3.0[e,f]
Extreme exposure[g]	7.5	7.0	6.0	6.0	5.5	5.0	4.5[e,f]	4.0[e,f]

[a]These quantities of mixing water are for use in computing cement factors for trial batches. They are maximal for reasonably well-shaped, angular coarse aggregates graded within limits of accepted specifications.

[b]The slump values for concrete containing aggregate larger than $1\frac{1}{2}$ in. are based on slump tests made after removal of particles larger than $1\frac{1}{2}$ in. by wet screening.

[c]These quantities of mixing water are for use in computing cement factors for trial batches when 3-in. or 6-in. nominal maximum-size aggregate is used. They are average for reasonably well-shaped coarse aggregates, well graded from coarse to fine.

[d]Additional recommendations for air content and necessary tolerances on air content for control in the field are given in a number of ACI documents, including ACI 201, 345, 318, 301, and 302. ASTM C94 for ready-mixed concrete also gives air content limits. The requirements in other documents may not always agree exactly; so in proportioning concrete, consideration must be given to selecting an air content that will meet the needs of the job and also meet the applicable specifications.

[e]For concrete containing large aggregates that will be wet screened over the $1\frac{1}{2}$-in. sieve prior to testing for air content, the percentage of air expected in the $1\frac{1}{2}$-in.-minus material should be tabulated in the $1\frac{1}{2}$-in. column. However, initial proportioning calculations should include the air content as a percent of the whole.

[f]When using large aggregate in low-cement-factor concrete, air entrainment need not be detrimental to strength. In most cases, the mixing water requirement is reduced sufficiently to improve the water/cement ratio and thus to compensate for the strength-reducing effect of entrained-air concrete. Generally, therefore, for these large maximum sizes of aggregate, air contents recommended for extreme exposure should be considered even though there may be little or no exposure to moisture and freezing.

[g]These values are based on the criterion that 9% air is needed in the mortar phase of the concrete. If the mortar volume will be substantially different from that determined in this recommended practice, it may be desirable to calculate the needed air content by taking 9% of the actual mortar volume.

Photo 14   North Shore Synagogue, Glencoe, Illinois. (Courtesy of Portland Cement Association.)

Flexural Strength of Plain Concrete Beams

This experiment is performed by three-point loading of plain concrete beams that are 6 by 6 by 18 in. in size, and that have spans three times their depth.

Tensile Splitting Tests

These tests are performed by loading the standard 6-in. by 12-in. cylinder by a line load perpendicular to its longitudinal axis. The tensile splitting strength can be defined as

$$f'_t = \frac{2P}{\pi DL}$$

where P = total value of the line load registered by the testing machine

D = diameter of the concrete cylinder

L = cylinder height

The results of all these tests give the designer a measure of the expected strength of the designed concrete in the built structure.

TABLE 2.7 RELATIONSHIP BETWEEN WATER/CEMENT
RATIO AND COMPRESSIVE STRENGTH OF CONCRETE

Compressive strength at 28 days[a] (psi)[b]	Water/cement ratio, by weight	
	Non-air-entrained concrete	Air-entrained concrete
6000	0.41	—
5000	0.48	0.40
4000	0.57	0.48
3000	0.68	0.59
2000	0.82	0.74

[a]Values are estimated average strengths for concrete containing not more than the percentage of air shown in Table 2.6. For a constant water/cement ratio, the strength of concrete is reduced as the air content is increased.

Strength is based on 6-in. × 12-in. cylinders moist-cured 28 days at 73.4 ± 3°F (23 ± 1.7°C) in accordance with Section 9(b) of ASTM C31, "Making and Curing Concrete Compression and Flexure Test Specimens in the Field.

Relationship assumes maximum size of aggregate about $\frac{3}{4}$ to 1 in.; for a given source, strength produced for a given water/cement ratio will increase as maximum size of aggregate decreases.

[b]1000 psi = 6.9 MPa.

TABLE 2.8 VOLUME OF COARSE AGGREGATE PER UNIT OF VOLUME
OF CONCRETE

Maximum size of aggregate (in.)[a]	Volume of dry-rodded coarse aggregate[b] per unit volume of concrete for different fineness moduli of sand			
	2.40	2.60	2.80	3.00
$\frac{3}{8}$	0.50	0.48	0.46	0.44
$\frac{1}{2}$	0.59	0.57	0.55	0.53
$\frac{3}{4}$	0.66	0.64	0.62	0.60
1	0.71	0.69	0.67	0.65
$1\frac{1}{2}$	0.75	0.73	0.71	0.69
2	0.78	0.76	0.74	0.72
3	0.82	0.80	0.78	0.76
6	0.87	0.85	0.83	0.81

[a]1 in. = 25.4 mm.

[b]Volumes are based on aggregates in dry-rodded condition as described in ASTM C29, "Unit Weight of Aggregate." These volumes are selected from empirical relationships to produce concrete with a degree of workability suitable for usual reinforced construction. For less workable concrete, such as that required for concrete pavement construction, they may be increased about 10%. For more workable concrete, the coarse aggregate content may be decreased up to 10%—provided that the slump and water/cement ratio requirements are satisfied.

Photo 15   (a) $4\frac{1}{2}$-in. slump mix; (b) $1\frac{1}{2}$-in. slump mix.

2.5 PLACING AND CURING OF CONCRETE

2.5.1 Placing

The techniques necessary for placing concrete depend on the type of member to be cast—whether it is a column, a beam, a wall, a slab, a foundation, a mass concrete dam, or an extension of previously placed and hardened concrete. For beams, columns and walls, the forms should be well oiled and thoroughly moistened to about a depth of 6 in. to avoid absorption of the moisture present in the wet concrete. Concrete should always be placed in horizontal layers that are compacted by means of high-frequency, power-driven vibrators of either the immersion or external type, as the case may be. It must be kept in mind, however, that overvibration may be harmful, for it could cause segregation of the aggregate and bleeding of the concrete.

2.5.2 Curing

Hydration of the cement takes place in the presence of moisture at temperatures above 50°F. It is necessary to maintain such a condition so that the chemical hydration reaction may occur. If drying is too rapid, a surface cracking termed plastic shrinkage takes place. It would result in the reduction of concrete strength due to cracking as well as the failure to attain full chemical hydration.

To facilitate good curing conditions, any of the following methods can be used:

1. Sprinkling continuously with water
2. Pondling with water
3. Covering the concrete with wet burlap, plastic film, or waterproof curing paper
4. Using liquid-membrane-forming curing compounds to retain the original moisture in the wet concrete
5. Steam curing in cases where the concrete member is manufactured under factory conditions, such as for precast beams, pipes, and prestressed girders and poles. Steam-curing temperatures are about 150°F. Curing time is usually 1 day compared to 5 to 7 days in other methods.

2.6 PROPERTIES OF HARDENED CONCRETE

The mechanical properties of hardened concrete can be classified as short-term or instantaneous properties and long-term properties. The short-term properties can be enumerated as (a) strength in compression, tension, and shear and (b) stiffness measured by modulus of elasticity. The long-term properties can be classified in terms of creep and shrinkage. The following sections present some details of the aforementioned properties.

2.6.1 Compressive Strength

Depending on the type of mix, the properties of aggregate, and the time and quality of the curing, compressive strengths of concrete up to 15,000 psi or more can be obtained. Commercial production of concrete with ordinary aggregate is usually in the 3000- to 10,000-psi range with the most common concrete strengths in the range of 3000 and 6000 psi.

The compressive strength, f_c', is based on standard 6-in. by 12-in. cylinders cured under standard laboratory conditions and tested at a specified rate of loading at 28 days of age. The standard specifications used in the United States are usually taken from ASTM C39. It should be mentioned that the strength of concrete in the actual structure may not be the same as that of the cylinder because of the difference in compaction and curing conditions. Photo 16 shows a typical cylinder in a compression-testing machine.

The ACI code specifies, for a strength test, the average of two cylinders from the same sample tested at the same age, which is usually 28 days. As for the frequency of testing, the code specifies that the strength level of an individual class of concrete can be considered satisfactory if (a) the average of all sets of three consecutive strength tests equals or exceeds the required f_c' and (b) no individual strength test (average of two cylinders) falls below the required f_c' by more than 500 psi.

The average concrete strength for which a concrete mix must be designed should exceed f_c' by an amount that depends on the uniformity of plant production and its prior documented record of test results.

2.6.2 Tensile Strength

The tensile strength of concrete is relatively low. A good approximation for the tensile strength f_{ct} is $0.10f_c' < f_{ct} < 0.20f_c'$. It is more difficult to measure tensile strength than compressive strength because of the gripping problems with testing machines. A number of methods are available for tension testing, with the most commonly used method being the cylinder splitting test. (See photo 17.)

For members subjected to bending, the value of the modulus of rupture f_r rather than tensile splitting strength f_t' is used in design. The modulus of rupture is measured by testing to failure plain concrete beams 6 in. square in cross section, having a span of 18 in. and loaded at third points (ASTM C78). The ACI specifies a value of $7.5\sqrt{f_c'}$ for the modulus of normalweight concrete.

Photo 16 Cylinder compression test.

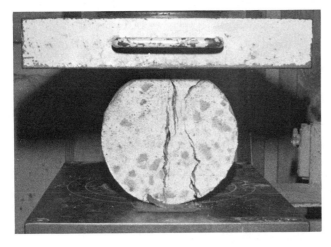

Photo 17 Tensile splitting test.

Photo 18   Concrete cylinders tested to failure in compression. Specimen A, low-epoxy-cement content; specimen B, high-epoxy-cement content. (Tests by Nawy, Sun, and Sauer.)

In most cases, lightweight concrete has a lower tensile strength than normalweight concrete. Here are the code stipulations for lightweight concrete.

i)   If the splitting tensile strength f_{ct} is specified, then

$$f_r = 1.09 f_{ct} \leq 7.5 \sqrt{f_c'}$$

ii)   If f_{ct} is not specified, a factor of 0.75 is used for all lightweight concrete and 0.85 is used for "sand-lightweight" concrete. Linear interpolation may be used for a mixture of natural sand and lightweight fine aggregate.

2.6.3 Stress–Strain Curve

A knowledge of the stress–strain relationship of concrete is essential in developing all analysis and design terms and procedures in concrete structures. Figure 2.2 shows typical stress–strain curves for various concrete strengths obtained from tests that used cylindrical concrete specimens loaded in uniaxial compression as reported by the Portland Cement Association. After approximately 70% of the failure stress, the material loses a large portion of its stiffness, thereby increasing the curvilinearity of the diagram. At the ultimate load, cracks parallel to the direction of loading become distinctly visible and most concrete cylinders except those with very low strengths fail

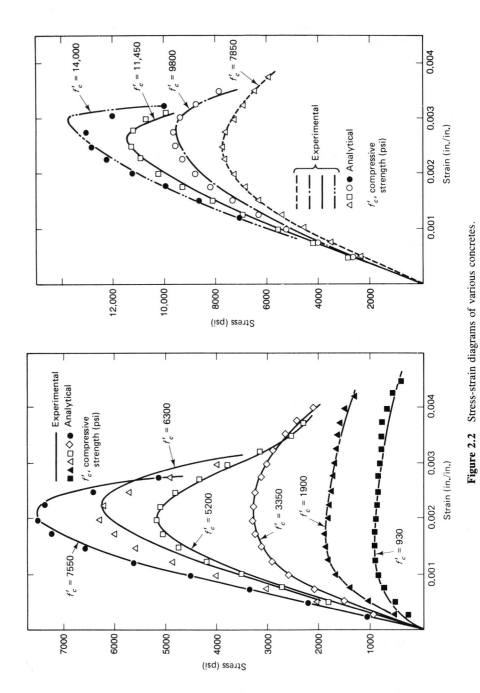

Figure 2.2 Stress-strain diagrams of various concretes.

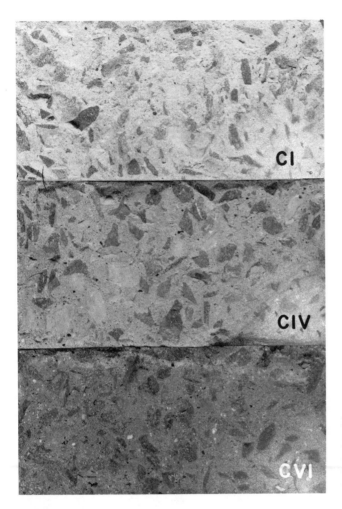

Photo 19 Fracture surfaces in tensile splitting tests of concretes with different *w/c* contents. Specimens CI and CIV with higher *w/c* content, hence more bond failures than specimen CVI. (Tests by Nawy, et al.)

suddenly shortly afterward. It should be noted that the higher the compressive strength of the concrete, the larger is the linear portion of this stress–strain diagram.

The modulus of elasticity of the concrete defining the slope of the tangent to the stress–strain diagram is

$$E_c = 33w^{1.5}\sqrt{f_c'} \tag{a}$$

where w is the unit weight of concrete in lbs per cu ft and f_c' is the cylinder compressive strength (psi). For normal weight concrete, where $w = 150$ pcf, the expression in Eq.(a) becomes

$$E_c = 57,000\sqrt{f_c'}$$

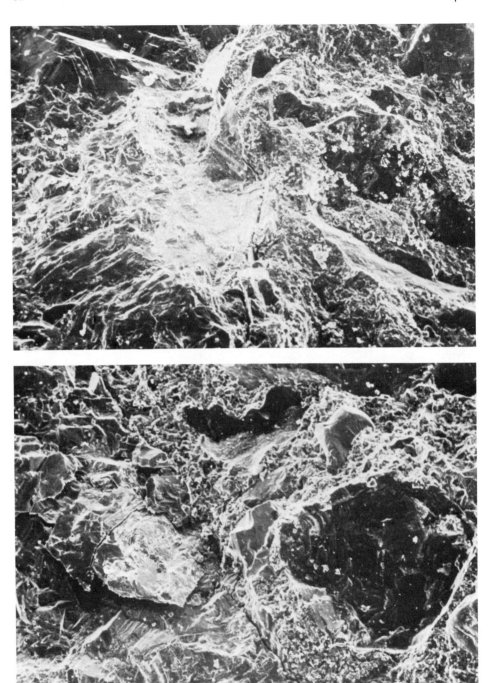

Photo 20 Electron microscope photographs of concrete from specimens A and B in photo 18. (Tests by Nawy, et al.)

2.7 SHRINKAGE AND CREEP

2.7.1 Shrinkage

Two types of shrinkage occur in concrete: plastic shrinkage and drying shrinkage. Plastic shrinkage occurs during the first few hours after placing the fresh concrete in the forms. Exposed surfaces are more easily affected by exposure to dry air because of their large contact surface. In such cases, moisture evaporates faster from the concrete surface and is replaced by the bleeding water from the lower layers of the concrete elements. Drying shrinkages develop after the concrete has already attained its final set and a good portion of the hydration chemical process in the cement gel is accomplished.

Shrinkage is not a completely reversible process. If a concrete unit is saturated with water after having fully shrunk, it will not expand to its original volume. The rate decreases with time, for older concretes are more resistant to stress and consequently undergo less shrinkage, so that the shrinkage strain becomes almost asymptotic with time.

Several factors affect the magnitude of drying shrinkage.

1. *Aggregate:* The aggregate acts to restrain the shrinkage of the cement paste; thus concretes with a high aggregate content are less vulnerable to shrinkage. In addition, the degree of restraint of a given concrete is determined by the properties of aggregates: those with high modulus of elasticity or with rough surfaces are more resistant to the shrinkage process.
2. *Water/cement ratio*: The higher the water/cement ratio, the higher are the shrinkage effects.
3. *Size of the concrete element*: Both the rate and the total magnitude of shrinkage decrease with an increase in the volume of the concrete element. The duration of shrinkage is longer for larger members, however, because more time is needed for the drying process to reach internal regions. One year may be needed for drying to commence at a depth of 10 in. from the exposed surface and 10 years may be necessary for drying to begin at 24 in. below the external surface.
4. *Type of Cement*: Rapid-hardening cement shrinks somewhat more than other types whereas shrinkage-compensating cements minimize or eliminate shrinkage cracking if used with restraining reinforcement.
5. *Admixtures*: This effect varies, depending on the type of admixture. An accelerator like calcium chloride that is used to accelerate the hardening and setting of the concrete increases the shrinkage. Pozzolans can also increase the drying shrinkage whereas air-entraining agents have little effect.
6. *Amount of reinforcement*: Reinforced concrete shrinks less than plain concrete; the relative difference is a function of the reinforcement percentage. Shrinkage strain versus time as shown in Fig. 2.3 indicates the stabilizing asymptotic behavior with time.

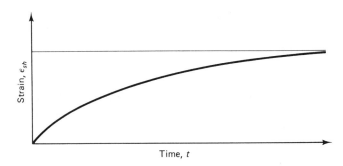

Figure 2.3 Shrinkage-time curve.

2.7.2 Creep

Creep or lateral flow is the increase in strain with time due to a sustained load. Initial deformation because of load is considered elastic strain; the additional strain due to the same sustained load is the creep strain.

Creep cannot be observed directly. It can only be determined by deducting elastic strain and shrinkage strain from the total deformation. Although shrinkage and creep are not independent phenomena, it can be assumed that superposition of strains is valid. Therefore

$$\text{Total strain } (\varepsilon_t) = \text{elastic strain } (\varepsilon_e) + \text{creep } (\varepsilon_c) + \text{shrinkage } (\varepsilon_{sh})$$

The composition of a concrete specimen can be essentially defined by the water/cement ratio, aggregate and cement types, and aggregate and cement contents. Therefore, like shrinkage, an increase in the water/cement ratio and in the cement content increases creep. Moreover, as in shrinkage, the aggregate induces a restraining effect such that an increase in aggregate content reduces creep.

Creep recovery versus time is schematically shown in Fig. 2.4. The residual strain seen in the plot causes a residual stress in the reinforcement, thereby increasing the stress level in the structure.

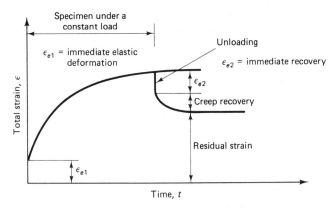

Figure 2.4 Creep recovery versus time.

2.8 REINFORCEMENT

2.8.1 General Properties

Concrete is strong in compression but weak in tension. Thus reinforcement is needed to resist the tensile stresses resulting from the induced loads. Additional reinforcement is occasionally used to reinforce the compression zone of concrete beam sections. Such steel is necessary for heavy loads in order to reduce long-term deflections.

Steel reinforcement for concrete consists of bars, wires, or welded wire fabric, all of which are manufactured in accordance with ASTM standards. The most important properties of reinforcing steel are as follows:

1. Young's modulus, E_s
2. Yield strength, f_y
3. Ultimate strength, f_u
4. Size or diameter of the bar or wire

Steel reinforcement is normally designated as grade 40, 60, and 80 steels. The steels have corresponding yield strengths of 40,000, 60,000, and 80,000 psi (276, 345, and 517 N/mm², respectively) and generally have a well-defined yield point. For steels that lack a well-defined yield point, the yield strength value is taken as the strength corresponding to a unit strain of 0.005 for grades 40 and 60 steels and 0.0035 for grade 80 steel. The ultimate tensile strengths corresponding to the 40, 60, and 80 grade steels are 70,000, 90,000, and 100,000 psi.

The percentage elongation at fracture varies with the grade, bar diameter, and manufacturing source, ranging from 4.5 to 12% over an 8-in. gage length. For most steels, the behavior is assumed to be elastoplastic and Young's modulus is taken as 29×10^6 psi.

To increase the bond between concrete and steel, projections called deformations are rolled on the bar surface (see Fig. 2.5) in accordance with ASTM specifications. The deformations shown must satisfy ASTM specifications A616-76 to be accepted as deformed bars. The deformed wire has indentations pressed into the wire or bar to serve as deformations. Except for wire used in spiral reinforcement in columns, only deformed bars, deformed wires, or wire fabric made from smooth or deformed wire may be used in reinforced concrete under approved practice.

Figure 2.6 shows typical stress–strain curves for grade 40, 60, and 80 steels.

Table 2.9 presents the reinforcement grade strengths and Table 2.10 gives the geometrical properties of the various-sized reinforcing bars.

Welded wire fabric is increasingly used for slabs because of the ease of placing the fabric sheets, control of reinforcement spacing, and better bond. The fabric reinforcement is made of smooth or deformed wires that run in perpendicular directions and are welded together at intersections. Table 2.11 presents the geometrical properties of some standard wire reinforcement.

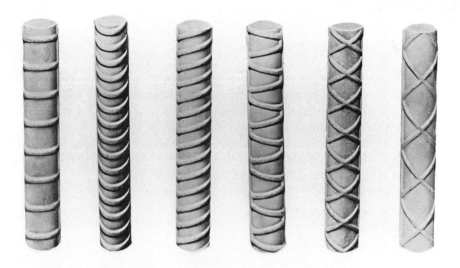

Figure 2.5   Various forms of ASTM-approved deformed bars.

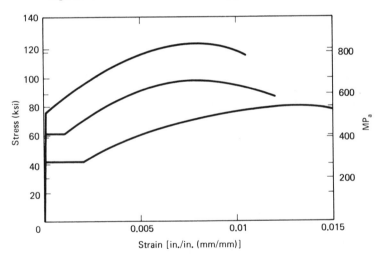

Figure 2.6   Typical stress-strain diagrams for various steels.

2.8.2 Bar Spacing and Concrete Cover for Steel Reinforcement

It is necessary to guard against honeycombing and to ensure that the wet concrete mix passes through the reinforcing steel without separation. Because the graded aggregate size in structural concrete often contains $\frac{3}{4}$-in. (19 mm diameter) coarse aggregate, minimum allowable bar spacing and minimum required cover as stipulated by ACI

TABLE 2.9 REINFORCEMENT GRADES AND STRENGTHS

1982 standard type	Minimum yield point or yield strength, f_y (psi)	Ultimate strength, f_u (psi)
Billet steel (A615)		
Grade 40	40,000	70,000
Grade 60	60,000	90,000
Axle steel (A617)		
Grade 40	40,000	70,000
Grade 60	60,000	90,000
Low-alloy steel (A706): Grade 60	60,000	80,000
Deformed wire		
Reinforced	75,000	85,000
Fabric	70,000	80,000
Smooth wire		
Reinforced	70,000	80,000
Fabric	65,000, 56,000	75,000, 70,000

TABLE 2.10 WEIGHT, AREA, AND PERIMETER OF INDIVIDUAL BARS

Bar Designation Number	Weight per foot (lb.)	1982 standard nominal dimensions		
		Diameter, d_b[in. (mm)]	Cross-sectional area, A_b (in.^{2})	Perimeter (in.)
3	0.376	0.375 (9)	0.11	1.178
4	0.668	0.500 (13)	0.20	1.571
5	1.043	0.625 (16)	0.31	1.963
6	1.502	0.750 (19)	0.44	2.356
7	2.044	0.875 (22)	0.60	2.749
8	2.670	1.000 (25)	0.79	3.142
9	3.400	1.128 (28)	1.00	3.544
10	4.303	1.270 (31)	1.27	3.990
11	5.313	1.410 (33)	1.56	4.430
14	7.65	1.693 (43)	2.25	5.32
18	13.60	2.257 (56)	4.00	7.09

codes are needed. Additionally, to protect the reinforcement from corrosion and loss of strength in case of fire, codes specify a minimum required concrete cover. Several major requirements of the ACI 318 code are listed here.

TABLE 2.11 STANDARD WIRE REINFORCEMENT

W&D size		U.S. customary			Area (in.²/ft of width for various spacings) Center-to-center spacing (in.)						
Smooth	Deformed	Nominal diameter (in.)	Nominal area (in.²)	Nominal weight (lb/ft)	2	3	4	6	8	10	12
W31	D31	0.628	0.310	1.054	1.86	1.24	0.93	0.62	0.465	0.372	0.31
W30	D30	0.618	0.300	1.020	1.80	1.20	0.90	0.60	0.45	0.366	0.30
W28	D28	0.597	0.280	0.952	1.68	1.12	0.84	0.56	0.42	0.336	0.28
W26	D26	0.575	0.260	0.934	1.56	1.04	0.78	0.52	0.39	0.312	0.26
W24	D24	0.553	0.240	0.816	1.44	0.96	0.72	0.48	0.36	0.288	0.24
W22	D22	0.529	0.220	0.748	1.32	0.88	0.66	0.44	0.33	0.264	0.22
W20	D20	0.504	0.200	0.680	1.20	0.80	0.60	0.40	0.30	0.24	0.20
W18	D18	0.478	0.180	0.612	1.08	0.72	0.54	0.36	0.27	0.216	0.18
W16	D16	0.451	0.160	0.544	0.96	0.64	0.48	0.32	0.24	0.192	0.16
W14	D14	0.422	0.140	0.476	0.84	0.56	0.42	0.28	0.21	0.168	0.14
W12	D12	0.390	0.120	0.408	0.72	0.48	0.36	0.24	0.18	0.144	0.12
W11	D11	0.374	0.110	0.374	0.66	0.44	0.33	0.22	0.165	0.132	0.11
W10.5		0.366	0.105	0.357	0.63	0.42	0.315	0.21	0.157	0.126	0.105
W10	D10	0.356	0.100	0.340	0.60	0.40	0.30	0.20	0.15	0.12	0.10
W9.5		0.348	0.095	0.323	0.57	0.38	0.285	0.19	0.142	0.114	0.095
W9	D9	0.338	0.090	0.306	0.54	0.36	0.27	0.18	0.135	0.108	0.09
W8.5		0.329	0.085	0.289	0.51	0.34	0.255	0.17	0.127	0.102	0.085
W8	D8	0.319	0.080	0.272	0.48	0.32	0.24	0.16	0.12	0.096	0.08
W7.5		0.309	0.075	0.255	0.45	0.30	0.225	0.15	0.112	0.09	0.075
W7	D7	0.298	0.070	0.238	0.42	0.28	0.21	0.14	0.105	0.084	0.07
W6.5		0.288	0.065	0.221	0.39	0.26	0.195	0.13	0.097	0.078	0.065
W6	D6	0.276	0.060	0.204	0.36	0.24	0.18	0.12	0.09	0.072	0.06
W5.5		0.264	0.055	0.187	0.33	0.22	0.165	0.11	0.082	0.066	0.055
W5	D5	0.252	0.050	0.170	0.30	0.20	0.15	0.10	0.075	0.06	0.05
W4.5		0.240	0.045	0.153	0.27	0.18	0.135	0.09	0.067	0.054	0.045
W4	D4	0.225	0.040	0.136	0.24	0.16	0.12	0.08	0.06	0.048	0.04
W3.5		0.211	0.035	0.119	0.21	0.14	0.105	0.07	0.052	0.042	0.035
W3		0.195	0.030	0.102	0.18	0.12	0.09	0.06	0.045	0.036	0.03
W2.9		0.192	0.029	0.098	0.174	0.116	0.087	0.058	0.043	0.035	0.029
W2.5		0.178	0.025	0.085	0.15	0.10	0.075	0.05	0.037	0.03	0.025
W2		0.159	0.020	0.068	0.12	0.08	0.06	0.04	0.03	0.024	0.02
W1.4		0.135	0.014	0.049	0.084	0.056	0.042	0.028	0.021	0.017	0.014

1. Clear distance between parallel bars in a layer must not be less than the bar diameter d_b or 1 in. (25.4 mm).
2. Clear distance between longitudinal bars in columns must not be less than $1.5d_b$ or 1.5 in. (38.1 mm).
3. Minimum clear cover in cast-in-place concrete beams and columns should not be less than 1.5 in. (38.1 mm) when there is no exposure to weather or contact with the ground; this same cover requirement also applies to stirrups, ties, and spirals.

In the case of slabs, plates, shells, and folded plates, where concrete is not exposed to a severe environment and where the reinforcement size does not exceed a No. 11 bar diameter (85.8 mm), the clear cover should not be less than $\frac{3}{4}$ in. (19 mm). Detailed requirements as to thickness of cover for various conditions can be found in various codes of practice, such as the Underwriters' National Building Code and the ACI code.

2.9 CONCRETE STRUCTURAL SYSTEMS

Architects and engineers design and build structures to serve a particular function. Form and function go hand in hand and the best structural system is the one that fulfills most needs of the user while being serviceable, attractive, and, hopefully, economically cost efficient. Although most structures are designed for a life span of 50 years, the durability performance record indicates that properly proportioned concrete structures have generally had longer useful lives.

Numerous concrete landmarks can be cited in which major credit is due to the art and science of structural design applied with ingenuity, logic, and imagination. Such concrete structural systems as the TWA Terminal, New York; the Newark Terminal, New Jersey; Symphony Hall, Melbourne, Australia; Chicago's Marina Towers and Water Tower Place; the Dallas Super Dome; and the Trump Towers in New York are a testimony to the mixture of form and function with superior engineering judgment. Photographs of several such landmarks appear throughout the book.

Such concrete systems consist of various concrete structural elements that, when synthesized, produce a total system. The components can be broadly classified into floor slabs, beams, columns, walls, and foundations.

Floor Slabs

Floor slabs are the main horizontal elements that transmit the moving live loads as well as the stationary dead loads to the vertical framing supports of a structure. They can be slabs on beams, as in Fig. 2.7, or waffle slabs, slabs without beams (flat plates) resting directly on columns, or composite slabs on joists. They can be proportioned so that they act in one direction (one-way slabs) or in two perpendicular directions (two-way slabs and flat plates).

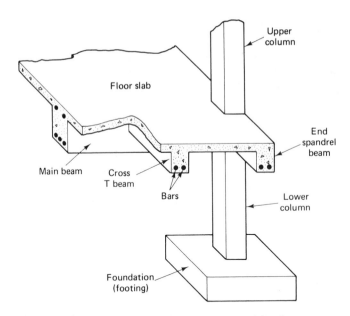

Figure 2.7   Typical reinforced concrete structural framing system.

Beams

Beams are the structural elements that transmit the tributory loads from floor slabs to vertical supporting columns. They are normally cast monolithically with the slab. They form a T-beam section for interior beams or an L beam at the building exterior, as seen in Fig. 2.7. The plan dimensions of a slab panel determine whether the floor slab behaves essentially as a one-way or a two-way slab.

Columns

The vertical elements support the structural floor system. They are compression members subjected, in most cases, to both bending and axial load and are of major importance in the safety considerations of any structure. If a structural system also consists of horizontal compression members, such members would be considered beam–columns.

Walls

Walls are the vertical enclosures for building frames. They are not usually or necessarily made of concrete but of any material that aesthetically fulfills the form and functional needs of the structural system. Additionally, structural concrete walls are often necessary as foundation walls, stairwell walls, and shear walls that resist horizontal wind loads and earthquake-induced loads.

Foundations

Foundations are the structural concrete elements that transmit the weight of the superstructure and the superimposed loads to the supporting soil. They could be in

Photo 21   LaGuardia Airport parking garage ramps. (Courtesy of Port Authority of New York and New Jersey.)

many forms, the simplest being the isolated footing shown in Fig. 2.7. It can be viewed as an inverted slab transmitting a distributed load from the soil to the column. Other forms of foundations are piles driven to rock; combined footings supporting more than one column; and mat foundations or rafts, which are basically inverted slab and beam construction.

In order to achieve a sound structure having the components just described, quality concrete and sound construction practices should be followed.

Quality concrete can be produced if adequate quality control and quality assurance are exercised in all stages of production and in the selection of all constituent materials. As the concrete is placed in the forms, the curing process must be fullly attained and the sequence of stripping the formwork and reshoring when necessary should be well planned and correctly executed.

Control tests to determine the compressive and tensile splitting strength should agree fully with ASTM standards and full loading of the finished system realized after the concrete has achieved its 28-day strength as a minimum. Transient loads during the construction process must be strictly controlled, for they can reach levels higher than actual design loads when the shored concrete can least sustain them. The recommendations given in this chapter, if followed, can result in quality concrete consistent with the environment it is expected to service.

The fabrication and placement of the reinforcement are other factors of major significance. The reinforcement should meet all minimum ASTM standards of strength, elongation, size, and condition. It should be accurately placed in the forms and well anchored in order to be able to contribute its full capacity. This topic is discussed further in Chapter 3 on flexural design of elements.

In short, all phases of formwork construction—mix proportioning, quality of the concrete components, the reinforcement, the placement process, curing of the plastic concrete, shoring and reshoring, and control of the construction loads—must be well monitored in order to comply with the structural design requirements discussed in the following chapters if the designed systems are to be safe and enduring.

SELECTED REFERENCES

2.1 American Society for Testing and Materials, "Annual Book of ASTM Standards—Part 14, Concrete and Mineral Aggregates," ASTM, Philadelphia, 1983, 834 pp.

2.2 ACI Committee 221, "Selection and Use of Aggregates for Concrete," *ACI Journal*, American Concrete Institute Proc. Vol. 58, No. 5, 1961, pp. 113–142.

2.3 American Concrete Institute, *ACI Manual of Concrete Practice 1983*, Part 5.

2.4 ACI Committee 212, "Admixtures for Concrete," *Manual of Concrete Practice*, Detroit, 1983, ACI 212.1 R81, 29 pp.

2.5 Nawy, E. G., M. M. Ukadike, and J. A. Sauer, "High-Strength Field-Modified Concretes," *Journal of the Structural Division*, ASCE, Vol. 103, No. ST12, December 1977, pp. 2307–2322.

2.6 American Concrete Institute, *Super-plasticizers in Concrete*, ACI Special Publication, SP62, Detroit, 1979, 427 pp.

2.7 Mindness, S., and J. F. Young, *Concrete*, Prentice-Hall, Englewood Cliffs, NJ, 1981, 671 pp.

2.8 Nawy, E. G., and P. N. Balaguru, "High Strength Concrete," Chapter 5, *Handbook of Structural Concrete*, Pitman Books, London, McGraw-Hill, New York, 1983, 1968 pp.

2.9 ACI Committee 211, "Standard Practice for Selecting Proportions for Normal and Mass Concrete (ACI 211.1–81)," American Concrete Institute, 1981, 32 pp.

2.10 ACI Committee 211, "Standard Practice for Selecting Proportions for Structural Lightweight Concrete (ACI 211.1–81)," American Concrete Institute, 1981, 18 pp.

2.11 Nawy, E. G., "Strength, Serviceability and Ductility," Chapter 12 in *Handbook of Structural Concrete*, McGraw-Hill, New York, 1983, pp. 12.1–12.88.

2.12 Nawy, E. G., *Reinforced Concrete—A Fundamental Approach*, Prentice-Hall Inc., Englewood Cliffs, NJ, 1985, 720 pp.

2.13 Portland Cement Association, *Design and Control of Concrete Mixtures*, 12th Ed., Skokie, Ill., 1979, 140 pp.

CHAPTER

3

Flexural Analysis and Design

3.1 INTRODUCTION

Loads acting on a structure, whether live gravity loads or other types, such as horizontal wind loads or those due to shrinkage and temperature, result in bending and deformation of the constituent structural elements. The bending of the beam element is the result of the deformational strain caused by the flexural stresses due to the external load.

As the load is increased, the beam sustains additional strain and deflection, leading to development of flexural cracks along the span of the beam. Continuous increases in the level of the load lead to failure of the structural element when the external load reaches the capacity of the element. Such a load level is termed the *limit state of failure in flexure*. Consequently, the designer must design the cross section of the element or beam so that it does not develop excessive cracking at service load levels and so that it has adequate safety and reserve strength to withstand the applied loads or stresses without failure.

Flexural stresses are a result of the external bending moments. They control, in most cases, the selection of the geometrical dimensions of a reinforced concrete section. The design process through the selection and analysis of a section is usually started by satisfying the flexural (bending) requirements except for special components, such as footings. Thereafter other factors, such as shear capacity, deflection, cracking, and bond development of the reinforcement, are analyzed and satisfied.

While the input data for the analysis of sections differ from the data needed for design, every design is essentially an analysis. One assumes the geometrical properties of a section in a design and proceeds to analyze such a section to determine if it can safely carry the required external loads. Thus a good understanding of the fundamental

43

Photo 22 Empire State Performing Arts Center, Albany, New York, during construction.

principles in the analysis procedure significantly simplifies the task of designing sections. The basic mechanics of materials principles of equilibrium of internal couples should be adhered to at all stages of loading.

If a beam is made up of homogeneous, isotropic, and linearly elastic material, the maximum bending stress can be obtained by using the well-known beam flexure formula, $f = Mc/I$. At ultimate load, the reinforced concrete beam is neither homogeneous nor elastic, thereby making that expression inapplicable for evaluating the stresses. However, the basic principles of the theory of bending can still be used to analyze reinforced concrete beam cross sections. Figure 3.1 shows a typical simply supported reinforced concrete beam. If the beam is so proportioned that all its constituent materials attain their capacity prior to failure, both the concrete and the steel fail simultaneously at midspan when the ultimate strength of the beam is reached. The corresponding strain and stress diagrams are shown in Fig. 3.2.

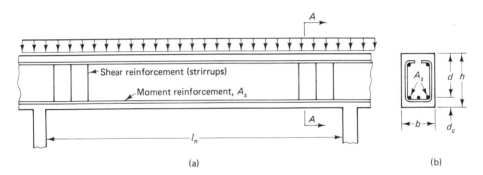

Figure 3.1   Typical reinforced concrete beam: (a) elevation; (b) section A-A.

The following assumptions are made in defining the behavior of the section.

1. Strain distribution is assumed to be linear. This assumption is based on Bernoulli's hypothesis that plane sections before bending remain plane and perpendicular to the neutral axis after bending.
2. Strain in the steel and the surrounding concrete is the same prior to cracking of the concrete or yielding of the steel.
3. Concrete is weak in tension. It cracks at an early stage of loading at about 10% of its limit compressive strength. Consequently, concrete in the tension zone of the section is neglected in the flexural analysis and design computations and the tension reinforcement is assumed to take the total tensile force.

To satisfy the equilibrium of the horizontal forces, the compressive force C in the concrete and the tensile force T in the steel should balance each other—that is,

$$C = T \qquad (3.1)$$

The terms in Fig. 3.2 are defined as follows:

b = width of the beam at the compression side

d = depth of the beam measured from the extreme compression fiber to the centroid of steel area

h = total depth of the beam

Photo 23   Simply-supported beam in flexural failure. (Tests by Nawy.)

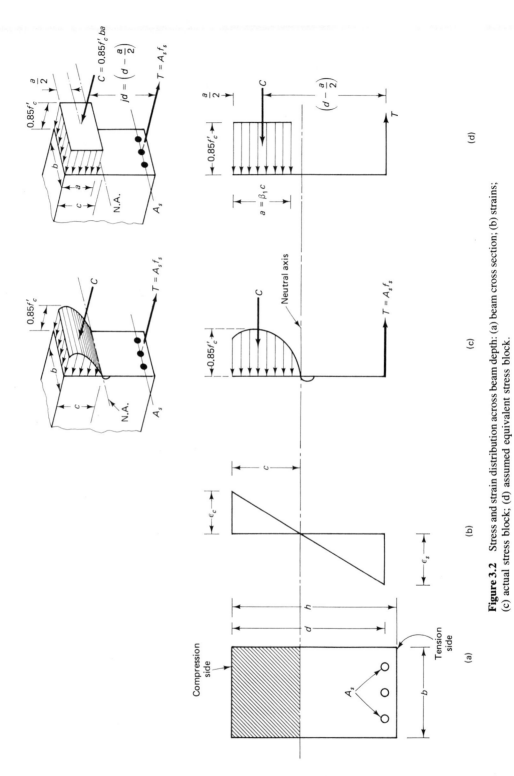

Figure 3.2 Stress and strain distribution across beam depth: (a) beam cross section; (b) strains; (c) actual stress block; (d) assumed equivalent stress block.

46

A_s = area of the tension steel

ϵ_c = strain in extreme compression fiber

ϵ_s = strain at the level of tension steel

f_c' = compressive strength of the concrete

f_s = stress in the tension steel

f_y = yield strength of the tension reinforcement

c = depth of the neutral axis measured from extreme compression fibers

3.2 ACI LOAD FACTORS AND SAFETY MARGINS

The ACI safety factors are termed *load factors,* for they restrict the estimation of reserve strength to the loads only. The estimated service or working loads are magnified by the coefficients, such as a coefficient of 1.4 for dead loads and 1.7 for live load. The types of normally occurring loads can be identified as (1) dead load, D; (2) live load, L; (3) wind load, W; (4) loads due to lateral pressure such as from soil in a retaining wall, H; (5) lateral fluid pressure loads, F; (6) loads due to earthquake, E; and (7) loads due to time-dependent effects, such as creep or shrinkage.

The basic combination of vertical loads is dead load plus live load. The *dead load,* which constitutes the weight of the structure and other relatively permanent features, can be estimated more accurately than the live load. The *live load* is estimated by using the weight of nonpermanent loads, such as people and furniture. The transient nature of live loads makes them difficult to estimate more accurately. Therefore a higher load factor is normally used for live loads than for dead loads. If the combination of loads consists only of live and dead loads, the ultimate load can be taken as

$$U = 1.4D + 1.7L \tag{i}$$

Structures are seldom subjected to dead and live loads alone; wind load is often present. For structures in which wind load should be considered, the recommended combination is

$$U = 0.75(1.4D + 1.7L + 1.7W) \tag{ii}$$

Maximum dead, live, and wind loads rarely, if ever, occur simultaneously. So the total factored load has to be reduced by using a reduction factor of 0.75. Because wind load is applied laterally, it is possible that the absence of vertical live load while wind load is present will produce maximum stress. The following load combination should also be used to arrive at the maximum value of the factored load U:

$$U = 0.9D + 1.3W \tag{iii}$$

Structures that must resist lateral pressure due to earth fill or fluid pressure should be designed for the worst of the following combination of factored loads:

$$U = 1.4D + 1.7L + 1.7H \tag{a}$$

$$U = 0.9D + 1.7H \tag{b}$$

$$U = 1.4D + 1.7L \tag{c}$$

$$U = 1.4D + 1.7L + 1.4F \tag{d}$$

$$U = 0.9D + 1.4F \tag{e}$$

$$U = 1.4D + 1.7L \tag{f}$$

The following combinations must be considered for earthquake loading:

$$U = 0.75(1.4D + 1.7L + 1.87E) \tag{g}$$

$$U = 0.9D + 1.43E \tag{h}$$

or $\qquad U \geq 1.4D + 1.7L \tag{i}$

whichever is largest. The philosophy used for combining the various load components for earthquake loading is essentially the same as that used for wind loading.

3.2.1 Design Strength Versus Nominal Strength: Strength Reduction Factor ϕ

The strength of a particular structural unit calculated by using the current established procedures is termed *nominal strength*. In the case of a beam, for example, the resisting moment capacity of the section calculated using the equations of equilibrium and the properties of concrete and steel is called the *nominal resisting moment strength M_n* of the section. This nominal strength is reduced by using a strength reduction factor, ϕ, to account for inaccuracies in construction, such as in the dimen-

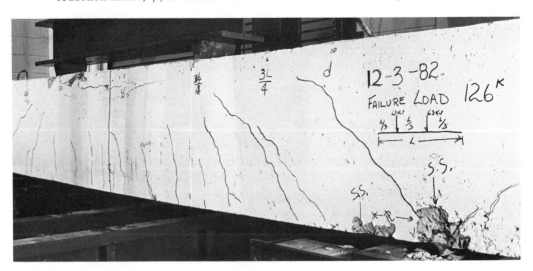

Photo 24   Flexural behavior of a reinforced concrete beam. (Test by Nawy, et al.)

sions or position of reinforcement or variations in properties. The reduced strength of the member is defined as the design strength of the member.

For a beam, the design moment strength ϕM_n should be at least equal to or slightly greater than the external factored moment M_u for the worst condition of factored load U. The factor ϕ varies for the different types of behavior and for the different types of structural elements. For beams in flexure, for instance, the reduction factor is 0.9.

For tied columns that carry dominant compressive loads, the ϕ factor equals 0.7. The smaller strength reduction factor used for columns is due to the structural importance of the columns in supporting the total structure compared to other members and to guard against progressive collapse and brittle failure with no advance warning of collapse. Beams, on the other hand, are designed to undergo excessive deflections before failure. Thus the inherent capability of the beam for advanced warning of failure permits the use of a higher strength reduction factor or resistance factor.

Table 3.1 summarizes the resistance factors ϕ for various structural elements as given in the ACI code.

TABLE 3.1   RESISTANCE OR STRENGTH
REDUCTION FACTOR ϕ

Structural element	Factor ϕ
Beam or slab: Bending or flexure	0.9
Columns with ties	0.7
Columns with spirals	0.75
Columns carrying very small axial loads (refer to Chapter 9 for more details)	0.7–0.9, or 0.75–0.9
Beam: Shear and torsion	0.85

3.3 THE EQUIVALENT RECTANGULAR BLOCK

The actual distribution of the compressive stress in a section has the form of a rising parabola, as shown in Fig. 3.2c. It is time consuming to evaluate the volume of the compressive stress block if it has a parabolic shape. An equivalent rectangular stress block due to Whitney can be used with ease and without loss of accuracy to calculate the compressive force and hence the flexural moment strength of the section. This equivalent stress block has a depth a and an average compressive strength $0.85f_c'$. As seen from Fig. 3.2d, the value of $a = \beta_1 c$ is determined by using a coefficient β_1 such that the area of the equivalent rectangular block is approximately the same as that of the parabolic compressive block, resulting in a compressive force C of essentially the same value in both cases.

The $0.85f_c'$ value for the average stress of the equivalent compressive block is based on the core test results of concrete in the structure at a minimum age of 28 days. Based on exhaustive experimental tests, a maximum allowable strain of 0.003 in./in. was adopted by the ACI as a safe limiting value. Even though several forms of stress blocks, including trapezoidal, have been proposed to date, the simplified equivalent rectangular block is accepted as the standard in the analysis and design of reinforced concrete. The behavior of the steel is assumed to be elastoplastic, as shown in Fig. 3.3a.

Using all the preceding assumptions, the stress distribution diagram shown in Fig. 3.2c can be redrawn as shown in Fig. 3.2d. One can easily deduce that the compression force C can be written $0.85f_c'\,ba$—that is, the *volume* of the compressive block at or near the ultimate when the tension steel has yielded, $\epsilon_s > \epsilon_y$. The tensile force T can be written $A_s f_y$. Thus equilibrium Eq. 3.1 can be rewritten

$$0.85f_c'ba = A_s f_y \tag{3.2}$$

or

$$a = \frac{A_s f_y}{0.85f_c'b} \tag{3.3}$$

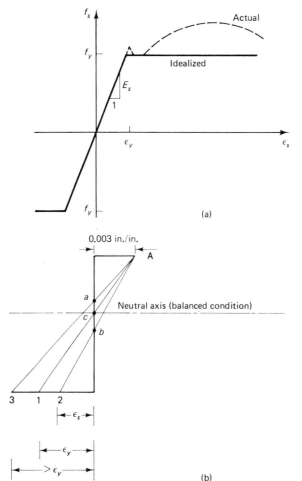

Figure 3.3 Strain distribution across depth: (a) idealized stress-strain diagram of the reinforcement; (b) strain distribution for various modes of flexural failure.

The moment of resistance of the section—that is, the nominal strength M_n—can be expressed as

$$M_n = (A_s f_y)jd \quad \text{or} \quad M_n = (0.85 f_c' ba)jd \tag{3.4a}$$

where jd is the lever arm, denoting the distance between the compression and tensile forces of the internal resisting couple. Using the simplified equivalent rectangular stress block from Fig. 3.2d, the lever arm is

$$jd = d - \frac{a}{2}$$

So the nominal moment of resistance becomes

$$M_n = A_s f_y \left(d - \frac{a}{2} \right) \tag{3.4b}$$

Because $C = T$, the moment equation can also be written

$$M_n = 0.85 f_c' ba \left(d - \frac{a}{2} \right) \tag{3.4c}$$

If the reinforcement ratio $\rho = A_s/bd$, Eq. 3.3 can be rewritten

$$a = \frac{\rho d f_y}{0.85 f_c'}$$

If $r = b/d$, Eq. 3.4c becomes

$$M_n = \rho r d^2 f_y \left(d - \frac{\rho d f_y}{1.7 f_c'} \right) \tag{3.5a}$$

Photo 25   Closeup of flexural cracks in photo 24.

or
$$M_n = [\omega r f_c'(1 - 0.59\omega)]d^3 \qquad (3.5b)$$

where $\omega = \rho f_y / f_c'$. Equation 3.5b is sometimes expressed as

$$M_n = Rbd^2 \qquad (3.6a)$$

where
$$R = \omega f_c'(1 - 0.59\omega) \qquad (3.6b)$$

Equations 3.5 and 3.6 are useful for the development of charts. A plot of the R value for singly reinforced beams is shown in Fig. 3.4.

If f_c', f_y, b, d, and A_s are given for a rectangular section and the beam is so proportioned and reinforced that failure occurs by simultaneous yielding of the tension

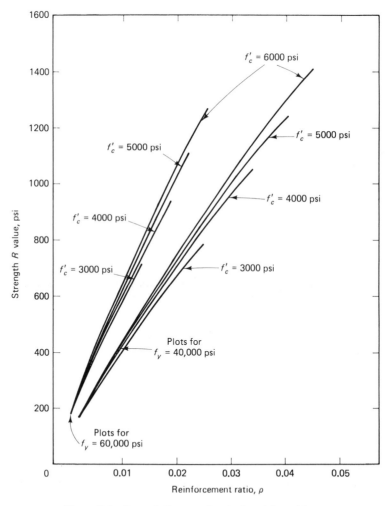

Figure 3.4 Strength-R curves for singly reinforced beams.

steel and crushing of the concrete at the compression side, the resisting moment strength can be obtained by using Eq. 3.4 or 3.5, but using the balanced steel area A_{sb} and the balanced rectangular block depth a_b instead of A_s and a. Beams, however, must be designed to fail in tension by initial yielding of the reinforcement, for reasons explained in subsequent sections.

Depending on the type of failure—that is, yielding of the steel or crushing of the concrete—three types of beams can be identified.

1. *Balanced section:*   The steel starts yielding when the concrete just reaches its ultimate strain capacity and commences to crush. At the start of failure, the permissible extreme fiber compressive strain is 0.003 in./in., whereas the tensile strain in the steel equals the yield strain $\epsilon_y = f_y/E_s$. The "balanced" strain distribution follows line $Ac\,1$ in Fig. 3.3b across the depth of the beam.

2. *Overreinforced section:*   Failure occurs by initial crushing of the concrete. At the initiation of failure, the steel strain ϵ_s will be lower than the yield strain ϵ_y, as in line $Ab\,2$, Fig. 3.3b; hence the steel stress f_s will be lower than its yield strength f_y. Such a condition is accomplished by using more reinforcement at the tension side than that required for the balanced condition.

3. *Underreinforced section:*   Failure occurs by initial yielding of the steel, as in line $Aa\,3$, Fig. 3.3b. The steel continues to stretch as the steel strain increases beyond ϵ_y. This condition is accomplished when the area of the tension reinforcement used in the beam is less than that required for the balanced strain condition.

It is to be noted from positions c, b, and a of the neutral axis that the axis rises toward the compressive fibers in the underreinforced beam as the limit state of failure

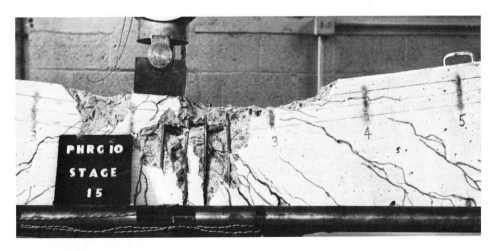

Photo 26 Beam at failure subjected to combined compression and bending. (Tests by Nawy, et al.)

Photo 27   Cracking level at rupture.

is reached. This behavior is easily identified in tests as the flexural cracks propagate toward the compression fibers until the concrete crushes. It should also be recognized that the vertical distances between points c, b, and a of the neutral axis from the extreme compression fibers for the three types of failure depend largely on the percentage ratio $\rho = A_s/bd$ but do not differ significantly because low values of strain are involved.

Concrete failure is sudden for it is a brittle material. Therefore almost all codes of practice recommend designing underreinforced beams to provide sufficient warning, such as excessive deflection before failure. In the case of statically indeterminate structures, ductile failure is essential for proper moment redistribution. Thus for beams, the ACI code limits the maximum amount of steel to 75% of that required for a balanced section. For practical purposes, however, the reinforcement ratio A_s/bd should not normally exceed 50%, to avoid congestion of the reinforcement and facilitate proper placing of the concrete. If the actual reinforcement ratio and the balanced reinforcement ratio are denoted as ρ and $\overline{\rho}_b$, respectively, then

$$\rho \leq 0.75\overline{\rho}_b \qquad (3.7a)$$

The code also stipulates the minimum steel requirement as

$$\rho > \frac{200}{f_y} \qquad (3.7b)$$

where f_y is expressed in psi, to account for temperature stresses and to ensure ductile failure in tension.

3.4 BALANCED REINFORCEMENT RATIO $\bar{\rho}_b$

To analyze a given beam, one must first determine the maximum allowable reinforcement ratio $0.75\bar{\rho}_b$. For rectangular sections reinforced only at the tension side, $\bar{\rho}_b$ is a function of only concrete strength and properties of steel—that is, modulus of elasticity E_s and yield strength f_y—irrespective of the section geometry. Using the strain distribution diagram in Fig. 3.2 for the balanced strain condition and from similar triangles, the relationship between the depth c (c_b for the balanced condition) of the neutral axis and the effective depth d can be written

$$\frac{c_b}{d} = \frac{0.003}{0.003 + f_y/E_s}$$

If E_s is taken as 29×10^6 psi, then

$$\frac{c_b}{d} = \frac{87,000}{87,000 + f_y} \tag{3.8a}$$

The relationship between the depth a of the equivalent rectangular stress block and the depth c of the neutral axis is

$$a = \beta_1 c \tag{3.8b}$$

The value of the stress block depth factor β_1 is as follows:

$$\beta_1 = \begin{cases} 0.85 & \text{for } 0 < f_c' \leq 4000 \text{ psi} \\ 0.85 - 0.05\left(\dfrac{f_c' - 4000}{1000}\right) & \text{for } 4000 \text{ psi} < f_c' \leq 8000 \text{ psi} \\ 0.65 & \text{for } f_c' > 8000 \text{ psi} \end{cases}$$

So for the balanced strain condition, the depth of the rectangular stress block is

$$a_b = \beta_1 c_b$$

For equilibrium of the horizontal forces,

$$A_{sb}f_y = 0.85f_c'ba_b$$

or

$$\bar{\rho}_b = \frac{A_{sb}}{bd} = \frac{0.85f_c'}{f_y}\frac{a_b}{d}$$

From Eq. 3.8 the balanced steel ratio becomes

$$\bar{\rho}_b = \beta_1\frac{0.85f_c'}{f_y}\frac{87,000}{87,000 + f_y} \tag{3.9}$$

where f_c' and f_y are expressed in psi. Thus if f_c' and f_y are known $\bar{\rho}_b$ and hence $0.75\bar{\rho}_b$ can be readily obtained regardless of the geometry of the concrete section.

Representative values of the maximum permissible reinforcement ratio ρ for singly reinforced beams are given in Table 3.2 both in pounds and in SI units. These values are 75% of the balanced reinforcement ratio $\overline{\rho}_b$ and should aid the reader in eliminating tedious computations of these frequently used values.

TABLE 3.2 MAXIMUM PERMISSIBLE REINFORCEMENT RATIO ($0.75\,\overline{\rho}_b \times 10^4$) FOR BEAMS WITH TENSION REINFORCEMENT ONLY (SINGLY REINFORCED BEAMS)[a]

f_y(psi)	$f_c' = 3000$ $\beta_1 = 0.85$	$f_c' = 4000$ $\beta_1 = 0.85$	$f_c' = 5000$ $\beta_1 = 0.80$	$f_c' = 6000$ $\beta_1 = 0.75$
40,000	278	371	437	491
50,000	206	275	324	364
60,000	160	214	252	283

[a]The values for the reinforcement ratio ρ in both the standard and SI units are almost identical since the ρ is a dimensionless quantity. ACI 318 M–83, for example, rounds up the conversion values such that for $f_y = 60,000$ psi $\simeq 400$ MPa and for $f_c' = 3000$ psi $\simeq 21$ MPa, $\rho = 160$, as above.

3.5 ANALYSIS OF SINGLY REINFORCED RECTANGULAR BEAMS FOR FLEXURE

The sequence of calculations presented in the flowchart of Fig. 3.5 can be used for the analysis of a given beam for both longhand and machine computations. The flowchart was developed by using the method of analysis presented in Section 3.2. The following examples illustrate typical analysis calculations following the flowchart logic in Fig. 3.5.

Photo 28   Beam subjected to combined axial load and bending. The neutral axis is at 70% of depth. (Tests by Nawy, et al.)

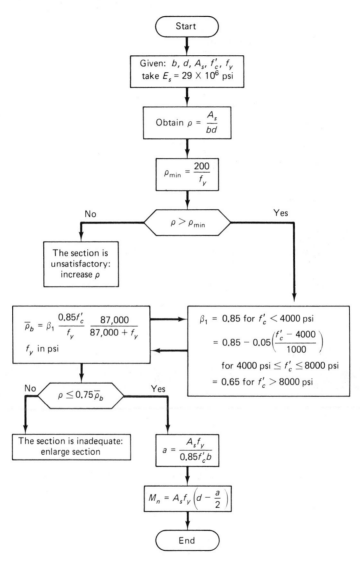

Figure 3.5   Flowchart for analysis of singly reinforced rectangular beams in bending.

Photo 29 Flexural cracking and deflection of beam subjected to flexure only prior to failure. (Tests by Nawy, et al.)

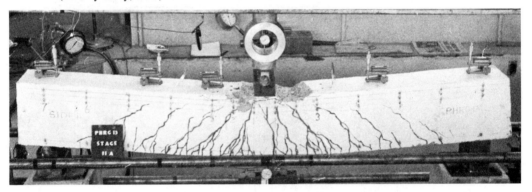

Photo 30 Crushing of concrete at compression side of beam subjected to flexure.

3.5.1 Example 3.1: Flexural Analysis of a Singly Reinforced Beam (Tension Reinforcement Only)

A singly reinforced concrete beam ($f_c' = 4000$ psi or 27.58 MPa) has the cross section shown in Fig. 3.6. Determine if the beam is overreinforced or underreinforced and if it satisfies the ACI code requirements for maximum and minimum reinforcement ratios for (a) $f_y = 60,000$ psi (413.4 MPa) and (b) $f_y = 40,000$ psi (275.6 MPa).

Solution

(a) From Eq. 3.9 for the balanced condition,

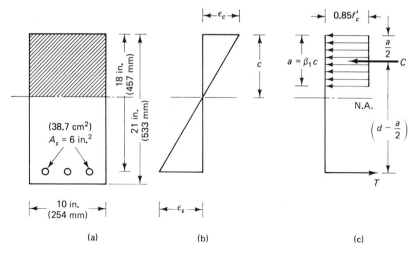

Figure 3.6 Stress and strain distribution in a typical singly reinforced rectangular section: (a) cross section; (b) strains; (c) stresses.

$$\bar{\rho}_b = \beta_1 \frac{0.85 f_c'}{f_y} \frac{87,000}{87,000 + f_y}$$

$$f_c' = 4000 \text{ psi}$$

$$f_y = 60,000 \text{ psi}$$

$$\beta_1 = 0.85 \qquad \text{for } f_c' = 4000 \text{ psi}$$

Therefore

$$\bar{\rho}_b = 0.85 \left(\frac{0.85 \times 4000}{60,000} \right) \frac{87,000}{87,000 + 60,000} = 0.029$$

$$A_{sb} = \bar{\rho}_b bd = 0.029 \times 10 \times 18$$

$$= 5.22 \text{ in.}^2 \ (3367 \text{ mm}^2) < A_s = 6 \text{ in.}^2 \ (3870 \text{ mm}^2)$$

$$\rho = \frac{6.0}{10 \times 18} = 0.033$$

Thus this section is overreinforced because $A_s > A_{sb}$ or $\rho > \bar{\rho}_b$ and does not satisfy ACI code requirements for ductility and maximum allowable reinforcement.

(b) In a new trial with $f_y = 40,000$ psi, one finds that

$$\bar{\rho}_b = 0.85 \left(\frac{0.85 \times 4000}{40,000} \right) \frac{87,000}{87,000 + 40,000} = 0.0495$$

$$A_{sb} = 8.91 \text{ in.}^2 \ (5746.95 \text{ mm}^2) > 6 \text{ in.}^2 \ (3367 \text{ mm}^2)$$

Therefore the section is underreinforced.

Minimum allowable reinforcement ratio $\rho_{min} = \dfrac{200}{f_y} = \dfrac{200}{40,000}$

$$= 0.005 << \text{actual } \rho$$

$$\text{Maximum allowable steel area} = 0.75 \times 8.91 = 6.68 \text{ in.}^2 > 6 \text{ in.}^2$$

Therefore the cross section satisfies ACI code requirements for maximum and minimum reinforcement. Note that the actual steel area in case (b) is only slightly less than the maximum allowable 75% of the balanced steel area. So congestion of steel is likely and the design can be improved by increasing the beam section size and reducing A_s.

3.5.2 Example 3.2: Nominal Resisting Moment in a Singly Reinforced Beam

For the beam cross section shown in Fig. 3.7, calculate the nominal moment strength if f_y is 60,000 psi (413.4 MPa) and f'_c is (a) 3000 psi (20.68 MPa), (b) 5000 psi (34.47 MPa), and (c) 9000 psi (62.10 MPa).

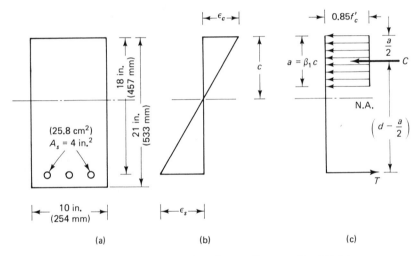

Figure 3.7 Beam cross-section strain and stress diagrams, Ex. 3.2: (a) cross section; (b) strains; (c) stresses.

Solution

$$b = 10 \text{ in. (254.0 mm)}$$

$$d = 18 \text{ in. (457.2 mm)}$$

$$A_s = 4 \text{ in.}^2 \text{ (2580 mm}^2\text{)}$$

$$f_y = 60,000 \text{ psi}$$

$$\rho_{min} = \frac{200}{f_y} = \frac{200}{60,000} = 0.003$$

$$\rho = \frac{A_s}{bd} = \frac{4}{10 \times 18} = 0.0222 > 0.003 \qquad \text{O.K.}$$

Note that f_y should be in psi units in the ρ_{min} expression.

(a) $f_c' = 3000$ psi (20.68 MPa).

$$\beta_1 = 0.85$$

Using Eq. 3.9, we have

$$\bar{\rho}_b = \beta_1 \frac{0.85 f_c'}{f_y} \frac{87,000}{87,000 + f_y}$$

$$= 0.85 \left(\frac{0.85 \times 3,000}{60,000} \right) \frac{87,000}{87,000 + 60,000} = 0.021$$

$$0.75 \bar{\rho}_b = 0.016$$

$$\rho > 0.75 \bar{\rho}_b$$

So the beam is considered overreinforced and does not satisfy the ACI requirements for ductility and maximum allowable reinforcement ratio.

(b) $f_c' = 5000$ psi (34.47 MPa).

$$\beta_1 = 0.85 - 0.05 \left(\frac{5000 - 4000}{1000} \right)$$

$$= 0.8$$

$$\bar{\rho}_b = 0.8 \left(\frac{0.85 \times 5000}{60,000} \right) \frac{87,000}{87,000 + 60,000}$$

$$= 0.034$$

$$0.75 \bar{\rho}_b = 0.025 > \rho = 0.0222 \qquad \text{O.K.}$$

$$A_s = 4 \text{ in.}^2$$

$$a = \frac{4 \times 60,000}{0.85 \times 5000 \times 10}$$

$$= 5.65 \text{ in.}$$

$$M_n = 4 \times 60,000 \left(18 - \frac{5.65}{2} \right) = 3,642,000 \text{ in.-lb}$$

$$= 303,500 \text{ ft-lb (411.52 kN-m)}$$

Or using Eq. 3.5 gives

$$\omega = \frac{0.0222 \times 60,000}{5000} = 0.267$$

$$M_n = \left[0.267 \times \frac{10}{18} \times 5000(1 - 0.59 \times 0.267)\right]18^3$$

$$= 3{,}644{,}019 \text{ in.-lb } (411.77 \text{ kN-m})$$

(c) $f_c' = 9000$ psi (62.10 MPa).

$$\beta_1 = 0.65 \text{ for } f_c' \geq 8000 \text{ psi}$$

$$\bar{\rho}_b = 0.65\left(\frac{0.85 \times 9000}{60{,}000}\right)\frac{87{,}000}{87{,}000 + 60{,}000}$$

$$= 0.049$$

$$0.75\,\bar{\rho}_b = 0.037 > \rho = 0.022 \qquad \text{O.K.}$$

$$a = \frac{4.0 \times 60{,}000}{0.85 \times 9000 \times 10} = 3.14 \text{ in. } (79.8 \text{ mm})$$

$$M_n = 4.0 \times 60{,}000\left(18 - \frac{3.14}{2}\right) = 3{,}943{,}200 \text{ in.-lb}$$

$$= 328{,}600 \text{ ft-lb } (445.6 \text{ kN-m})$$

3.5.3 Trial-and-Adjustment Procedures for the Design of Singly Reinforced Beams

In Ex. 3.2 the geometrical properties of the beam—that is, b, d, and A_s—were given. In a design example an assumption of width b (or the ratio b to d) and the level of reinforcement ratio ρ must be made. The ratio b/d varies between 0.25 and 0.6 in usual practice. Although the ACI code permits a reinforcement ratio ρ up to $0.75\,\bar{\rho}_b$, a ratio of $0.5\,\bar{\rho}_b$ is advisable to prevent congestion of steel, secure a good bond between the reinforcement and the adjacent concrete, and provide good deflection control.

Studies on cost optimum design indicate that cost-effective sections can be obtained by using a minimum practical b/d ratio and a maximum practical reinforcement ratio ρ within the above-stated limitations. So one could use the following steps to design the beam cross section by following the flowchart logic in Fig. 3.8.

1. Calculate the external factored moment. To obtain the beam self-weight, an assumption has to be made for the value of d. The minimum depth for deflection specified in the ACI code can be used as a guide. Assume a b/d ratio r between 0.25 and 0.6 and calculate $b = rd$. A first trial assumption $b \simeq d/2$ is recommended.
2. Choose a reinforcement ratio of approximately $0.5\,\bar{\rho}_b$.
3. (a) Select a value of moment factor R based on an assumed ρ value $\simeq 0.5\,\bar{\rho}_b$. Assuming that $b \simeq d/2$, calculate d for $M_n = Rbd^2$ and proceed to analyze the section.
   (b) Alternatively, choose d on the basis of minimum deflection requirement.

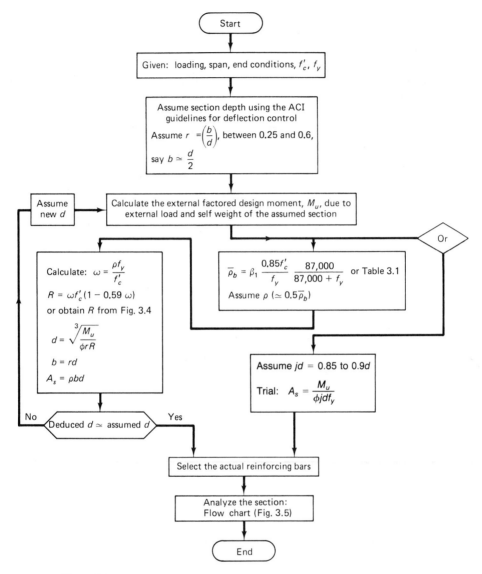

Figure 3.8 Flowchart for sequence of operations for the design of singly reinforced rectangular sections.

Choose a width b as in 3(a). assume a moment arm $jd \simeq 0.85d$ to $0.90d$. Calculate A_s as a first trial; then analyze the section, using $b = d/2$.

The process of arriving at the final section is highly convergent even by longhand computations in that it should not require more than three trial cycles. The use of handheld or desktop personal computers enormously simplifies the design–analysis process and permits the student or engineer to proportion sections at a fraction of the time needed when using handbooks, charts, or longhand computations, easy as these other means can be.

For designers who prefer charts, Eq. 3.6 ($M_n = Rbd^2$) can be used for the first trial in design. The value of R can be obtained from charts (see Fig. 3.4) for various values of ρ, f_c', and f_y available in handbooks.

3.5.4 Example 3.3: Design of a Singly Reinforced Simply Supported Beam for Flexure

A reinforced concrete simply supported beam has a span of 30 ft (9.14 m) and is subjected to a service uniform load $w_L = 1500$ lb/ft (21.9 kN/m), as shown in Fig. 3.9. Design a beam section to resist the factored external bending load. Given:

$f_c' = 4000$ psi (27.58 MPa)

$f_y = 60,000$ psi (413.4 MPa)

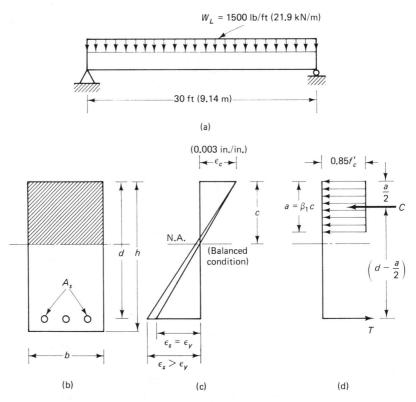

Figure 3.9 Simply-supported reinforced concrete uniformly-loaded beam: (a) elevation; (b) cross section; (c) strains; (d) stresses.

Solution

Assume a minimum thickness from the ACI code deflection table:

$$\frac{I_n}{16} = \frac{30 \times 12}{16} = 22.5 \text{ in.}$$

In order to estimate the preliminary self-weight, assume total thickness $h = 24.0$ in., effective depth $d = 20$ in., and width of the beam $b = 10$ in. ($r = b/d = 0.5$).

$$\text{Beam self-weight} = \frac{24 \times 10}{144} \times 150 = 250 \text{ lb/ft}$$

Factored load $W_u = 1.4D + 1.7L = 1.4 \times 250 + 1.7 \times 1500 = 2900$ lb/ft

$$\text{Required moment } M_u = \frac{w_u l_n^2}{8} = \frac{2900 \times 30^2}{8} \times 12 = 3{,}915{,}000 \text{ in.-lb}$$

$$\text{Required nominal resisting moment } M_n = \frac{M_u}{\phi} = \frac{3{,}915{,}000}{0.9} = 4{,}350{,}000 \text{ in.-lb}$$

Get $\bar{\rho}_b$ from Table 3.1, which gives $0.75\,\bar{\rho}_b$, or calculate:

$$\bar{\rho}_b = \beta_1 \frac{0.85 f_c'}{f_y} \frac{87{,}000}{87{,}000 + f_y} = 0.85 \left(\frac{0.85 \times 4000}{60{,}000} \right) \frac{87{,}000}{87{,}000 + 60{,}000} = 0.0285$$

Assume a reinforcement ratio $\rho = 0.5\,\bar{\rho}_b = 0.0143$.

$$\omega = \frac{\rho f_y}{f_c'} = \frac{0.0143 \times 60{,}000}{4000} = 0.215$$

Using Eq. 3.6b yields

$$R = \omega f_c'(1 - 0.59\omega) = 0.215 \times 4000(1 - 0.59 \times 0.215)$$

$$\approx 750$$

The value of R can also be obtained from the chart in Fig. 3.4, using the chosen ρ and the given values for f_c' and f_y.

Using Eq. 3.6a, one has $M_n = Rbd^2$ and assuming that $b = 0.5d$,

$$d = \sqrt[3]{\frac{M_n}{0.5R}} = \sqrt[3]{\frac{4{,}350{,}000}{0.5 \times 750}} = 22.64 \text{ in.}$$

$$b = 0.5 \times 22.64 = 11.32 \text{ in.}$$

Based on practical considerations, try a section with $b = 12$ in., $d = 23$ in., and $h = 26$ in.

$$\text{Revised self-weight} = \frac{12 \times 26}{144} \times 150 = 325 \text{ lb/ft}$$

Factored load $W_u = 1.4 \times 325 + 1.7 \times 1500 = 3005$ lb/ft

$$\text{Factored moment } M_u = \frac{3005(30)^2}{8} \times 12 = 4{,}056{,}750 \text{ in.-lb}$$

Required resisting moment $M_n = \dfrac{M_u}{\phi} = \dfrac{4,056,750}{0.9} = 4,507,500$ in.-lb

$A_s = \rho bd = 0.0143 \times 12 \times 23 = 3.95$ in.^{2}

Try three No. 10 bars (32.3 mm diameter) with $A_s = 3.81$ in.^{2}

$$\rho = \frac{3.81}{12 \times 23} = 0.0138 < 0.75\,\bar{\rho}_b > \rho_{min} \quad \text{O.K.}$$

Check the nominal moment strength of the assumed section:

$$a = \frac{A_s f_y}{0.85 f_c' b} = \frac{3.81 \times 60,000}{0.85 \times 4000 \times 12} = 5.60 \text{ in.}$$

$$M_n = 3.81 \times 60,000\left(23 - \frac{5.60}{2}\right) = 4,617,720 \text{ in.-lb (521.8 kN-m)}$$

$$> \text{required } M_n = 4,507,500 \text{ in.-lb}$$

Adopt the section. Note that the designed section resists a slightly larger moment than the required moment:

$$\text{Percent overdesign} = \frac{4,617,720 - 4,507,500}{4,507,500} = 2.45\%$$

which is a reasonable level expected in proportioning concrete elements. It is always necessary also to check that the web width can accommodate the number of bars in each layer based on concrete cover and minimum spacing requirements. In this example the minimum web width to accommodate three No. 10 bars $= 10.5$ in. $< b = 12.0$ in., which is O.K.

Alternative Solution by Trial and Adjustment

Assume that moment arm $jd \simeq 0.85d$:

$$\text{Minimum thickness } h = \frac{l_n}{16} = \frac{30 \times 12}{16} = 22.5 \text{ in.}$$

Try $h = 26$ in. (660.4 mm), $d = 23.0$ in. (584.2 mm), and $b \simeq \frac{1}{2}d \simeq 12$ in. (304.8 mm).

$$\text{Self-weight} = \frac{12 \times 26}{144} \times 150 = 325.0 \text{ lb/ft}$$

Factored load $U = 1.4 \times 325.0 + 1.7 \times 1500 = 3005$ lb/ft

$$\text{Factored moment } M_u = \frac{3005(30.0)^2}{8} \times 12$$

$$= 4,056,750 \text{ in.-lb (458.1 kN-m)}$$

$$\text{Required nominal resisting moment } M_n = \frac{M_u}{\phi} = \frac{4,056,750}{0.9}$$

$$= 4,507,500 \text{ in.-lb (509.3 kN-m)}$$

Moment arm $jd \simeq 0.85d \simeq 0.85 \times 23.0 = 19.55$ in.

$$M_n = A_s f_y \left(d - \frac{a}{2} \right) = A_s f_y jd \text{ or } 4,507,500 = A_s \times 60,000 \times 19.55$$

So

$$A_s = \frac{4,507,500}{60,000 \times 19.55} = 3.84 \text{ in.}^2$$

Try three No. 10 bars (32.3 mm diameter = 3.81 in.² (2457.5 mm²). Continue the design, following the flowchart in Fig. 3.8.

3.5.5 Arrangement of Reinforcement

Figure 3.10 shows the cross section of the beam at midspan. In arranging the reinforcing bars, one should satisfy the minimum cover requirements explained in Chapter 2. The required clear cover for beams is 1.5 in. (38 mm).

The stirrups shown in Fig. 3.10 should be designed to satisfy the shear requirements of the beam explained in Chapter 5. Two bars called *hangers* are placed on the compression side to support the stirrups. Reinforcement detailing provisions and bar development length requirements are discussed in Chapter 4.

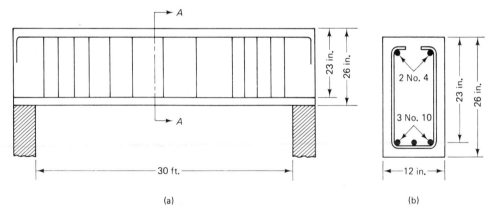

Figure 3.10 Details of reinforcement, Ex. 3.3: (a) sectional elevation (not to scale); (b) midspan section *A-A*.

3.6 ONE-WAY SLABS

One-way slabs are concrete structural floor panels for which the ratio of the long span to the short span equals or exceeds a value of 2.0. When this ratio is less than 2.0, the floor panel becomes a two-way slab or plate, as discussed later in this chapter. A one-way slab is designed as a singly reinforced 12-in. (304.8-mm)-wide beam strip using the same design and analysis procedure discussed earlier for singly reinforced beams. Figure 3.11 shows a one-way slab floor system.

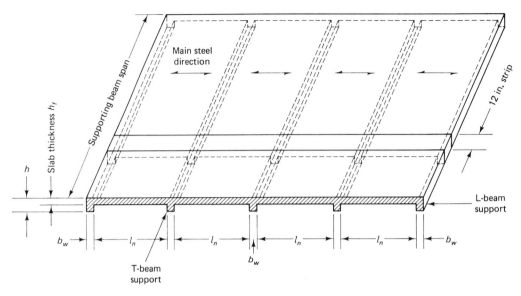

Figure 3.11 Isometric view of four-span continuous one-way-slab floor system.

Loading for slabs is normally specified in pounds per square foot (psf). One has to distribute the reinforcement over the 12-in. strip and specify the center-to-center spacing of the reinforcing bars. In slab design a thickness is normally assumed and the reinforcement is calculated by using a lever arm $0.9d$ for a first trial and then adjusting A_s for a lever arm $(d - a/2)$.

Supported slabs—namely, slabs not on grade—do not normally require shear reinforcement for typical loads. Transverse reinforcement has to be provided perpendicular to the direction of bending in order to resist shrinkage and temperature stresses. Shrinkage and temperature reinforcement should not be less than 0.002 times the gross area for grade 40 or 50 bars and 0.0018 for grade 60 steel and welded fire fabric.

3.6.1 Example 3.4: Design of a One-Way Slab for Flexure

A one-way single-span reinforced concrete slab has a simple span of 10 ft (3.05 m) and carries a live load of 120 psf (5.75 kPa) and a dead load of 20 psf (0.96 kPa) in addition to its self-weight. Design the slab and the size and spacing of the reinforcement at midspan, assuming a simple support moment. Given:

$f_c' = 4000$ psi (27.5 MPa), normalweight concrete

$f_y = 60,000$ psi (413.4 MPa)

Minimum thickness for deflection $= l/20$

Solution

$$\text{Minimum depth for deflection, } h = \frac{l}{20} = \frac{10 \times 12}{20} = 6 \text{ in. (152.4 mm)}$$

Assume for flexure an effective depth $d = 5$ in. (127 mm).

$$\text{Self-weight of a 12-in. strip} = \frac{6 \times 12}{144} \times 150 = 75 \text{ lb/ft (3.59 kN/m)}$$

Therefore

$$\text{Factored external load } U = 1.7 \times 120 + 1.4(20 + 75) = 337 \text{ lb/ft}$$

$$\text{Factored external moment } M_u = \frac{337 \times 10^2}{8} \times 12 \text{ in.-lb}$$

$$= 50{,}550 \text{ in.-lb (5712.15 kN-m)}$$

Assume that the arm $(d - a/2) = 0.9d = 0.9 \times 5 = 4.50$ in.

$$M_u = \phi A_s f_y \left(d - \frac{a}{2} \right)$$

Therefore

$$50{,}550 = 0.9 \times A_s \times 60{,}000(4.50)$$

or $A_s = 0.21$ in.^{2} per 12 in. of slab.

Trial-and-adjustment check for assumed moment arm:

$$a = \frac{A_s f_y}{0.85 f'_c b} = \frac{0.21 \times 60{,}000}{0.85 \times 4000 \times 12} = 0.31 \text{ in.   (8.6 mm)}$$

$$50{,}550 = 0.9 \times A_s \times 60{,}000 \left(5 - \frac{0.31}{2} \right)$$

$$A_s = 0.193 \text{ in.}^2 \text{ per 12-in. slab strip}$$

Use No. 4 bars at 12 in. center-to-center spacing (13-mm-diameter bars at 304.8 mm center to center) with an area of 0.20 in.^{2} or No. 3 bars at $6\frac{1}{2}$ in. center-to-center spacing.

$$\rho = \frac{0.20}{5.0 \times 12} = 0.0033 \qquad \rho_{\min} = \frac{200}{60{,}000} = 0.0033 = \rho \qquad \text{O.K.}$$

$$\rho_{\max} = 0.75\bar{\rho}_b = 0.75 \left(\frac{0.85 \times 4{,}000}{60{,}000} \times 0.85 \times \frac{87{,}000}{87{,}000 + 60{,}000} \right)$$

$$= 0.0214 > \rho \qquad \text{O.K.}$$

Shrinkage and temperature reinforcement:

$$\rho = 0.0018$$

$$\text{Area of steel} = 0.0018 \times 6 \times 12 = 0.13 \text{ in.}^2 = \text{No. 4 bars at 18 in. c/c}$$

Provide No. 4 bars at 18 in. center to center (maximum allowable spacing = 5h = 5 × 6 = 30 in.).

Thus this design can be adopted with slab thickness h = 6 in. (152.4 mm) and effecive depth d = 6.0 − (0.75 + 0.25) = 5 in. (127.0 mm) to satisfy the $\frac{3}{4}$-in. minimum concrete cover requirement. Use for main reinforcement No. 4 bars at 12 in. center to center and for temperature reinforcement No. 4 bars at 18 in. center to center, as shown in Fig. 3.12.

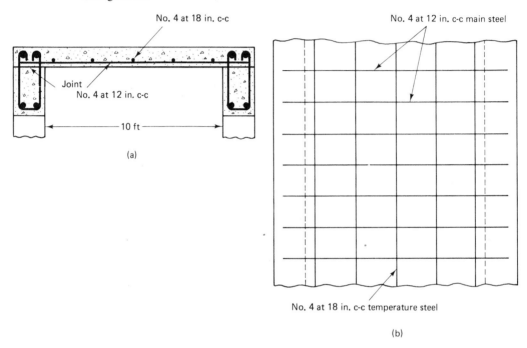

Figure 3.12   Reinforcement details of the one-way slab in Ex. 3.4: (a) sectional elevation; (b) reinforcement plan.

3.7 DOUBLY REINFORCED BEAMS

Doubly reinforced sections contain reinforcement both at the tension and at the compression face, usually at the support section only. They become necessary when either architectural limitations restrict the beam web depth at midspan or the midspan section is not adequate to carry the support negative moment even when the tensile steel at the support is sufficiently increased. In such cases, most bottom bars at midspan are extended and well anchored at the supports to act as compression reinforcement. The bar development length should be well established and the compressive and tensile steel at the support section well tied with closed stirrups to prevent buckling of the compressive bars.

In analysis or design of beams with compression reinforcement A_s', the analysis is so divided that the section is theoretically split into two parts, as shown in Fig. 3.13.

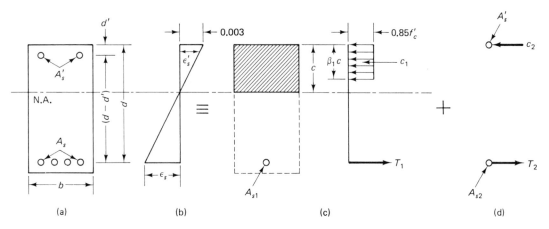

Figure 3.13 Doubly-reinforced beam design: (a) cross section; (b) strains; (c) part 1 of the solution—singly reinforced part; (d) part 2 of the solution—contribution of compression reinforcement.

The two parts of the solution constitute (a) the singly reinforced part involving the equivalent rectangular block, as discussed in Section 3.2, with the area of tension reinforcement being $(A_s - A_s')$; and (b) the two areas of equivalent steel A_s' at both the tension and compression sides to form the couple T_2 and C_2 as the second part of the solution.

It can be seen from Fig. 3.13 that the total nominal resisting moment $M_n = M_{n1} + M_{n2}$—that is, the summation of the moments for parts 1 and 2 of the solution.

Part 1

The tension force $T_1 = A_{s1}f_y = C_1$. But $A_{s1} = A_s - A_s'$ because equilibrium requires that A_{s2} at the tension side be balanced by an equivalent A_s' at the compression side. So the nominal resisting moment strength

$$M_{n1} = A_{s1}f_y\left(d - \frac{a}{2}\right) \quad \text{or} \quad M_{n1} = (A_s - A_s')f_y\left(d - \frac{a}{2}\right) \qquad (3.10a)$$

where

$$a = \frac{A_{s1}f_y}{0.85f_c'b} = \frac{(A_s - A_s')f_y}{0.85f_c'b}$$

Part 2

$$A_s' = A_{s2} = (A_s - A_{s1})$$
$$T_2 = C_2 = A_{s2}f_y$$

Taking the moment about the tension steel,

$$M_{n2} = A_{s2}f_y(d - d') \qquad (3.10b)$$

Adding the moments for parts 1 and 2 yields

$$M_n = M_{n1} + M_{n2} = (A_s - A_s')f_y\left(d - \frac{a}{2}\right) + A_s'f_y(d - d') \qquad (3.11a)$$

The design moment strength ϕM_n must be equal to or greater than the external factored moment M_u such that

$$M_u = \phi\left[(A_s - A_s')f_y\left(d - \frac{a}{2}\right) + A_s'f_y(d - d')\right] \qquad (3.11b)$$

This equation is valid *only* if A_s' yields. Otherwise the beam must be treated as a singly reinforced beam neglecting the compression steel or one must find the actual stress f_s' in the compression reinforcement A_s' and use the actual force in the moment equilibrium equation.

3.7.1 Strain-Compatibility Check

It is always necessary to verify that the strains across the depth of the section follow the linear distribution indicated in Fig. 3.13. In other words, a check is necessary to ensure that strains are compatible across the depth at the strength design levels. Such a verification is called a *strain-compatibility check*.

For A_s' to yield, the strain ϵ_s' in the compression steel should be greater than or equal to the yield strain of reinforcing steel, which is f_y/E_s. The strain ϵ_s' can be calculated from similar triangles. Referring to Fig. 3.13b, one has

$$\epsilon_s' = \frac{0.003(c - d')}{c}$$

or

$$\epsilon_s' = 0.003\left(1 - \frac{d'}{c}\right)$$

Because

$$c = \frac{a}{\beta_1} = \frac{(A_s - A_s')f_y}{\beta_1 \times 0.85f_c'b} = \frac{(\rho - \rho')f_y d}{\beta_1 \times 0.85f_c'}$$

then

$$\epsilon_s' = 0.003\left[1 - \frac{0.85\beta_1 f_c'd'}{(\rho - \rho')df_y}\right] \qquad (3.12)$$

As noted, for compression steel to yield, the following condition must be satisfied:

$$\epsilon_s' \geq \frac{f_y}{E_s} \quad \text{or} \quad \epsilon_s' \geq \frac{f_y}{29 \times 10^6}$$

The compression steel yields if

$$0.003\left[1 - \frac{0.85\beta_1 f_c'}{(\rho - \rho')f_y}\frac{d'}{d}\right] \geq \frac{f_y}{29 \times 10^6} \qquad (3.13)$$

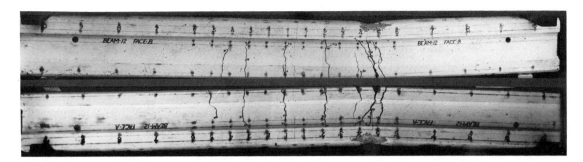

Photo 31 Flexural cracking in lightly reinforced beam. (Tests by Nawy, et al.)

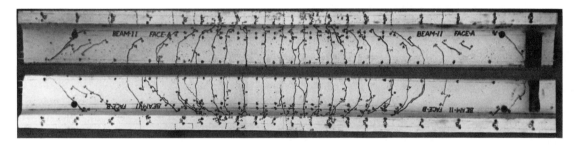

Photo 32 Flexural cracking in heavily reinforced beam. (Tests by Nawy, et al.)

or
$$-\frac{0.85\beta_1 f_c' d'}{(\rho - \rho')f_y d} \geq \frac{f_y - 87,000}{87,000}$$

or
$$\rho - \rho' \geq \frac{0.85\beta_1 f_c' d'}{f_y d}\frac{87,000}{87,000 - f_y} \qquad (3.14)$$

If ϵ_s' is less than ϵ_y, the stress in the compression steel, f_s', can be calculated as

$$f_s' = E_s \epsilon_s' = 29 \times 10^6 \epsilon_s' \qquad (3.15)$$

Using Eqs. 3.12 and 3.15 yields

$$f_s' = 29 \times 10^6 \times 0.003\left[1 - \frac{0.85\beta_1 f_c' d'}{(\rho - \rho')f_y d}\right] \qquad (3.16)$$

This value of f_s' can be used as a first approximation in the strain compatibility check in cases where the compression reinforcement did *not* yield. The reinforcement ratio for the balanced section can be written

$$\rho_b = \overline{\rho}_b + \rho'\frac{f_s'}{f_y} \qquad (3.17a)$$

where $\overline{\rho}_b$ corresponds to the balanced steel ratio for a singly reinforced beam that has a tension steel area A_{s1}.

The singly reinforced part of the solution in a doubly reinforced section normally uses the maximum allowable reinforcement ratio, $0.75\rho_b$. Consequently, the maximum allowable reinforcement ratio for a doubly reinforced beam can be expressed as

$$\rho \leq 0.75\overline{\rho}_b + \rho'\frac{f_s'}{f_y} \qquad (3.17b)$$

In this discussion adjustment for the concrete area replaced by the compression reinforcement is disregarded as being insignificant for practical design purposes. It is to be noted that in cases where the compression reinforcement A_s' did not yield, the depth of the rectangular compressive block should be calculated by using the actual stress in the compression steel from the calculated strain value ϵ_s' at the compression reinforcement level so that

$$a = \frac{A_sf_y - A_s'f_s'}{0.85f_c'b} \qquad (3.18)$$

Equation 3.16 can be used for the f_s' value in the first trial in order to obtain an "a" value and hence the first trial neutral axis depth value c. Once c is known, ϵ_s' can be evaluated from similar triangles in Fig. 3.13b, thereby obtaining the first approximation of f_s' to be used in recalculating a more refined value. More than one or two additional trials for calculating f_s' are not justified because undue refinement has negligible practical effect on the true value of the nominal moment strength M_n.

The nominal moment strength in Eq. 3.11 becomes in this case

$$M_n = (A_sf_y - A_s'f_s')\left(d - \frac{a}{2}\right) + A_s'f_s'(d - d') \qquad (3.19)$$

The flowchart (Fig. 3.14) can be used for the sequence of calculations in the analysis of doubly reinforced beams. Examples 3.5 and 3.6 illustrate the analysis and design of doubly reinforced sections.

3.7.2 Example 3.5: Analysis of a Doubly Reinforced Beam for Flexure

Calculate the nominal moment strength M_n of the doubly reinforced section shown in Fig. 3.15. Given:

$$f_c' = 5000 \text{ psi } (34.46 \text{ MPa}), \text{ normalweight concrete}$$

$$f_y = 60,000 \text{ psi } (413.4 \text{ MPa})$$

$$d' = 2.5 \text{ in. } (63.5 \text{ mm})$$

Solution

$$A_s = 5.08 \text{ in.}^2 \qquad \rho = \frac{A_s}{bd} = \frac{5.08}{14 \times 21} = 0.0173$$

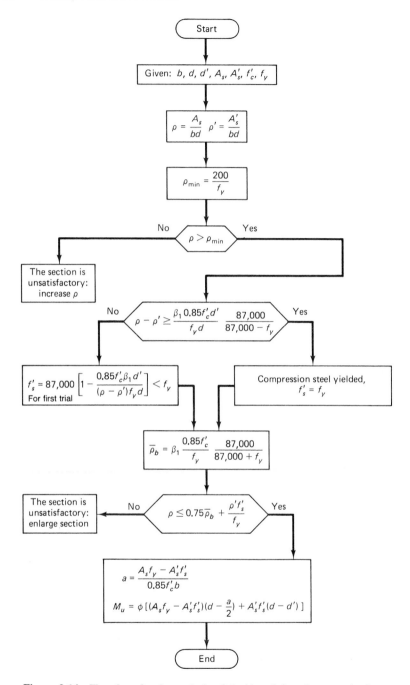

Figure 3.14 Flowchart for the analysis of doubly reinforced rectangular beam.

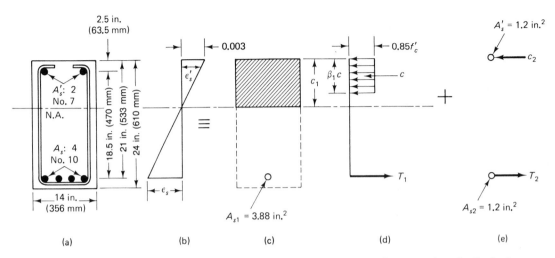

Figure 3.15 Doubly reinforced cross-section geometry and stress and strain distribution: (a) cross section; (b) strains; (c) part 1 section; (d) part 1 forces; (e) part 2 forces.

$$A_s' = 1.2 \text{ in.}^2 \qquad \rho' = \frac{A_s'}{bd} = \frac{1.2}{14 \times 21} = 0.0041$$

$$A_s - A_s' = A_{s1} = 5.08 - 1.2 = 3.88 \text{ in.}^2$$

$$\rho - \rho' = 0.0173 - 0.0041 = 0.0132$$

To check whether the compression steel has yielded, use Eq. 3.14:

$$\rho - \rho' \geq \frac{0.85\beta_1 f_c' d'}{f_y d} \frac{87,000}{87,000 - f_y}$$

$$\geq \frac{0.85 \times 0.80 \times 5000 \times 2.5}{60,000 \times 21} \frac{87,000}{87,000 - 60,000}$$

$$\geq 0.0217$$

The actual $(\rho - \rho') = 0.0132 < 0.0217$. Therefore the compression steel did not yield and f_s' is less than f_y. For the first trial in cases where the compression steel did not yield

$$f_s' = 87,000\left[1 - \frac{0.85\beta_1 f_c'}{(\rho - \rho')f_y} \frac{d'}{d}\right]$$

$$= 87,000\left(1 - \frac{0.85 \times 0.80 \times 5000}{0.0132 \times 60,000} \times \frac{2.5}{21}\right) = 42,538 \text{ psi.}$$

$$a = \frac{A_s f_y - A_s' f_s'}{0.85 f_c' b} = \frac{5.08 \times 60,000 - 1.2 \times 42,538}{0.85 \times 5000 \times 14} = 4.26 \text{ in. (108.33 mm)}$$

$$\text{Neutral axis depth } c = \frac{4.26}{0.80} = 5.325 \text{ in.}$$

From similar triangles in Fig. 3.15b the strain ϵ'_s at the compression steel level $= 0.0019$ in./in., giving $f'_s = 0.0019 \times 29 \times 10^6 = 46,155$ psi. An additional trial cycle for a more refined value of $a = 4.21$ in., hence $c = 5.26$ in., *gives* $f'_s = 45,650$ psi (314.76 kN).

$$\overline{\rho}_b = \beta_1 \frac{0.85\, f'_c}{f_y} \frac{87,000}{87,000 + f_y}$$

$$= 0.8 \frac{0.85 \times 5000}{60,000} \frac{87,000}{87,000 + 60,000} = 0.0335$$

So $0.75\,\overline{\rho}_b = 0.0252$

From Eq. 3.12 the maximum allowable reinforcement ratio

$$\rho \leq 0.75\,\overline{\rho}_b + \rho'\frac{f'_s}{f_y}$$

$$0.75\,\overline{\rho}_b + \rho'\frac{f'_s}{f_y} = 0.0252 + 0.0041 \times \frac{45,650}{60,000}$$

$$= 0.0283 > \rho = 0.0173 \qquad \text{O.K.}$$

$$a = \frac{5.08 \times 60,000 - 1.2 \times 45,650}{0.85 \times 5000 \times 14} = 4.20 \text{ in. } (106.73 \text{ mm})$$

$$M_n = (A_s f_y - A'_s f'_s)\left(d - \frac{a}{2}\right) + A'_s f'_s (d - d')$$

$$M_n = (5.08 \times 60,000 - 1.2 \times 45,650)\left(21.0 - \frac{4.20}{2}\right)$$

$$+ 1.2 \times 45,650(21.0 - 2.5) = 5,738,808 \text{ in.-lb } (648.49 \text{ kN-m})$$

Note that if the first trial value of $f'_s = 42,538$ psi from Eq. 3.16 is used, $M_n = 5,732,689$ in.-lb, which differs by less than 1% from the final M_n value. Such a low percentage difference can be justified by using the value of f'_s obtained from Eq. 3.16 for all practical purposes without resort to additional trials.

$$M_u = \phi M_n = 0.9 \times 5,738,808 = 5,164,927 \text{ in.-lb } (583.64 \text{ kN-m})$$

3.7.3 Trial-and-Adjustment Procedure for the Design of Doubly Reinforced Sections for Flexure

1. *Midspan section:* The trial-and-adjustment procedure described in Section 3.5 is followed in order to design the section at midspan if it is a rectangular section; otherwise follow the same procedure as that for the design of T beams and L beams (Section 3.8).

2. *Support section:* Width b and effective depth d are already known from part 1 together with the value of the external negative factored moment M_u.

    (a) Find the strength M_{n1} of a singly reinforced section, using the already established b and d dimensions of the section at midspan and a reinforcement ratio $\rho \leq 0.75\bar{\rho}_b$.

(b) From step (a) find $M_{n2} = M_n - M_{n1}$ and determine the resulting $A_{s2} = A'_s$. The total steel area at the tension side would be $A_s = A_{s1} + A'_s$.

(c) *Alternatively*, determine how many bars are extended from the midspan to the support to give the A'_s to be used in calculating M_{n2}.

(d) From step (c) find the value of $M_{n1} = M_n - M_{n2}$. Calculate A_{s1} for a singly reinforced beam as the first part of the solution. Then determine total $A_s = A_{s1} + A'_s$. Verify that A_{s1} does not exceed $0.75\bar{\rho}_b$ if it is revised in the solution.

(e) Check for the compatibility of strain in both alternatives to verify whether the compression steel yielded or not and use the corresponding stress in the steel for calculating the forces and moments.

(f) Check for satisfactory minimum reinforcement requirements.

(g) Select the appropriate bar sizes.

If it is necessary to design a doubly reinforced rectangular precast continuous beam, alternative method 3(a) or 3(b) of Section 3.5 for singly reinforced beams can be followed. An assumption is made of an R value higher than the R value that is used for singly reinforced beams for selection of the first trial section. Because it is not advisable to use an A'_s value larger than $\frac{1}{3}A_s$ or $\frac{1}{2}A_s$, assume that $R' \simeq 1.3$ to $1.5R$.

3.7.4 Example 3.6: Design of a Doubly Reinforced Beam for Flexure

A doubly reinforced concrete beam section has a maximum effective depth $d = 25$ in. (635 mm) and is subjected to a total factored moment $M_u = 9.4 \times 10^6$ in.-lb (1062 kN-m), including its self-weight. Design the section and select the appropriate reinforcement at the tension and the compression faces to carry the required load. Given:

$f'_c = 4000$ psi (27.58 MPa)

$f_y = 60,000$ psi (413.4 MPa)

Minimum effective cover $d' = 2.5$ in. (63.5 mm)

Solution

Assume that $b = 14$ in. $\simeq 0.55d$.

$$\bar{\rho}_b = \beta_1 \frac{0.85f'_c}{f_y} \frac{87,000}{87,000 + 60,000} = 0.85\left(\frac{0.85 \times 4000}{60,000}\right)\frac{87,000}{87,000 + 60,000}$$

$$= 0.0285$$

Or obtain $0.75\bar{\rho}_b$ from Table 3.1. Assume a tension reinforcement ratio 0.016 ($\simeq 0.5\bar{\rho}_b$) for the simply reinforced part of the solution.

Tension reinforcement area $A_{s1} = (A_s - A'_s) = 0.016 \times 14 \times 25 = 5.6$ in.^{2}

The resisting strength of a singly reinforced section of dimensions 14 in. × 25 in. and a tensile steel area $A_{s1} = 5.6$ in.^{2} is

$$M_{n1} = 5.6 \times 60,000\left(25.0 - \frac{5.6 \times 60,000}{2 \times 0.85 \times 4000 \times 14}\right)$$

$$= 7,214,118 \text{ in.-lb } (815.2 \text{ kN-m})$$

ϕM_{n1} is less than the total factored $M_u = 9.4 \times 10^6$ in.-lb; that is, the section is too small to support the required factored moment M_u. Therefore the section should be designed as doubly reinforced. The resisting moment corresponding to the singly reinforced part

$$M_{n1} = 7,214,118 \text{ in.-lb}$$

The moment to be resisted by the doubly reinforced part is

$$\frac{9,400,000}{0.9} - 7,214,118 = 3,230,326 \text{ in.-lb } (365 \text{ kN-m})$$

To verify if the compression steel A_s' has yielded, check

$$\rho - \rho' \geq \frac{0.85 f_c' \beta_1 d'}{f_y d} \quad \frac{87,000}{87,000 - f_y}$$

$$\geq \frac{0.85 \times 4000 \times 0.85 \times 2.5}{60,000 \times 25} \quad \frac{87,000}{87,000 - 60,000}$$

$$\geq 0.0155$$

The actual $\rho - \rho' = 0.016 > 0.0155$; so compression steel yielded

$$f_s' = f_y$$

Because $M_{n2} = A_s' f_y (d - d')$, $3,230,326 = A_s' \times 60,000(25.0 - 2.5)$, to give $A_s' = 2.39$ in.^{2}, corresponding to A_{s2}.

$$A_s = A_{s1} + A_s' = 5.6 + 2.39 = 7.99 \text{ in.}^2$$

Use eight No. 9 (28.6-mm-diameter) bars in *two layers* at the tension side ($A_s = 8.0$ in.^{2}) and four No. 7 (22.2-mm-diameter) bars in one layer at the compression side ($A_s' = 2.4$ in.^{2}), as shown in Fig. 3.16. Check if compression steel yielded in the final design:

$$\rho = \frac{8.0}{14 \times 25} = 0.02286 \qquad \rho' = \frac{2.4}{14 \times 25} = 0.00686$$

$$\rho - \rho' = 0.0160 < 0.75 \bar{\rho}_b \qquad \text{O.K.}$$

$$\rho - \rho' > 0.0155 \qquad \text{Hence } f_s' = f_y. \qquad \text{O.K.}$$

$$\text{Minimum reinforcement ratio} = \frac{200}{f_y} = \frac{200}{60,000} = 0.0033 < \rho \qquad \text{O.K.}$$

$$A_s - A_s' = 8.00 - 2.4 = 5.6 \text{ in.}^2$$

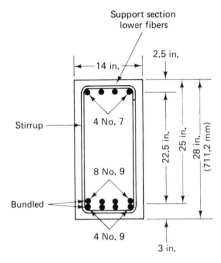

Support section
lower fibers

14 in.

2.5 in.

Stirrup

4 No. 7

8 No. 9

Bundled

4 No. 9

3 in.

22.5 in.

25 in.

28 in.

(711.2 mm)

Figure 3.16 Reinforcing details of the doubly reinforced beam in Ex. 3.6.

Design moment $M_u = 0.9 \left[5.6 \times 60,000 \left(25.0 - \dfrac{5.6 \times 60,000}{2 \times 0.85 \times 4,000 \times 14} \right) \right.$

$$\left. + 2.4 \times 60,000(25.0 - 2.5) \right]$$

$$= 9,408,706 \text{ in.-lb} > 9,400,000 \text{ in.-lb } (1063 \text{ kN-m} > 1062 \text{ kN-m})$$

Adopt the design.

Alternative Solution

Assume an R' value $\simeq 1.5R \simeq 1350$ (larger than $R = 900$ for singly reinforced beams of the same material properties).

$$M_n = \frac{9.4 \times 10^6}{0.9} = 10.44 \times 10^6 \text{ lb-in.}$$

$$M_n = Rbd^2 \quad \text{or} \quad 10.44 \times 10^6 = 1350bd^2$$

$$bd^2 = \frac{10.44 \times 10^6}{1350} = 7737 \text{ in.}^3$$

Assume that $b \simeq \frac{1}{2}d$; hence $d^3 = 15,474$ in.^{3} and $d = 24.92$ in. Assume a trial section with $b = 14$ in., $d = 25$ in. and proceed to analyze the section in the usual manner, first choosing $\rho - \rho' \leq \bar{\rho}_b$, as given in the preceding alternative solution.

3.8 NONRECTANGULAR SECTIONS (T and L Beams)

T beams and L beams are the most commonly used flanged sections. Because slabs are cast monolithically with the beams as shown in Fig. 3.17, additional stiffness or strength is added to the rectangular beam section from participation of the slab. Based

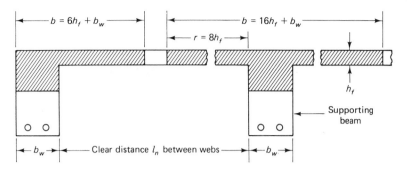

Figure 3.17   T and L beams as part of a slab beam floor system (cross section at beam midspan).

on extensive tests and longstanding engineering practice, a segment of the slab can be considered to act as a monolithic part of the beam across the beam flange. It should be noted that, in the case of composite sections, if the beam and slab are continuously shored during construction (supported continuously), they can be assumed to act together in supporting all loads, including their self-weight. If the beam is not shored, however, it must carry its weight plus the weight of the slab while it hardens. After the slab has hardened, the two together will support the additional loads.

The flange width accepted for inclusion with the beam in forming the flanged section must satisfy the following requirements:

T beams:

Effective overhand $\not> 8h_f$

Overhand width on each side $\not> \frac{1}{2}$ the clear distance to the face of the next web ($\not> \frac{1}{2} l_n$)

Flange width $b \not> \frac{1}{4}$ of supporting beam span $= \frac{1}{4}L$

Spandrel or edge beams (beams with a slab on one side only):

The effective overhang $\not> 6h_f$ nor $\not> \frac{1}{2}$ the clear distance to the next web ($\not> \frac{1}{2} l_n$) nor $\frac{1}{12}$ the span length of the beam.

Beams with overhang on one side are called *L beams.*

3.8.1 Analysis of T and L Beams

Flanged beams are considered primarily for use as sections at midspans, as shown in Fig. 3.17. This use occurs because the flange is in compression at midspan and can contribute to the moment strength of the midspan section. The flange is in tension at the support; consequently, it is disregarded in the flexural strength computations of the support section. In other words, the support section would be an inverted doubly reinforced section having the compressive steel A_s' at the bottom fibers and tensile steel A_s at the top fibers. Figure 3.18 shows an elevation of a continuous beam with sections taken at midspan and at the supports to illustrate this discussion.

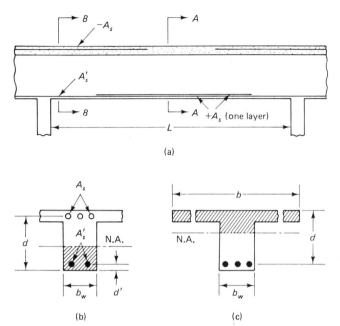

(a)

(b)   (c)

Figure 3.18 Elevation and sections of a monolithic continuous beam: (a) beam elevation; (b) support section *B-B* (inverted doubly reinforced beam); (c) midspan section *A-A* (real T beam).

Photo 33 Structural behavior of simply supported prestressed flanged beam. (Tests by Nawy, et al.)

The basic principles used for the design of rectangular beams are also valid for the flanged beams. The major difference between the rectangular and flanged sections is in the calculation of compressive force C_c. Depending on the depth of the neutral axis, c, the following cases can be identified.

Case 1: Depth of Neutral Axis c Less Than Flange Thickness h_f (Fig. 3.19)

This case can be treated similarly to the standard rectangular section, provided that the depth a of the equivalent rectangular block is less than the flange thickness. The flange width b of the compression side *should be used* as the beam width in the analysis.

Referring to Fig. 3.19 for force equilibrium, where C is equal to T, gives us

$$0.85f_c'ba = A_s f_y \quad or \quad a = \frac{A_s f_y}{0.85f_c' b}$$

The nominal moment strength would thus be $M_n = A_s f_y(d - a/2)$. This expression is the same as that of Eq. 3.4 for the rectangular section. Because the force contribution of concrete in the tension zone is neglected, it does not matter whether part of the flange is in the tension zone.

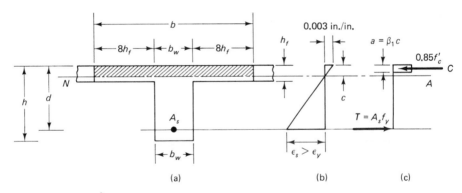

Figure 3.19   T-beam section with neutral axis within the flange $(c < h_f)$: (a) cross section; (b) strains; (c) stresses.

Case 2: Depth of Neutral Axis c Larger Than Flange Thickness h_f (Fig. 3.20)

Here $(c > h_f)$ the depth of the equivalent rectangular stress block a could be smaller or larger than the flange thickness h_f. If c is greater than h_f and a is less than h_f, the beam could still be considered as a rectangular beam for design purposes. So the design procedure explained for case 1 is applicable to this case.

If both c and a are greater than h_f, the section must be considered a T section. This type of T beam $(a > h_f)$ can be treated in a manner similar to that for a doubly reinforced rectangular cross section (Fig. 3.20). The contribution of the flange over-hang compressive force is considered analogous to the contribution of imaginary compressive reinforcement. In Fig. 3.20 the compressive force C_n is equal to the average concrete strength f_c' multiplied by the cross-sectional area of the flange overhangs.

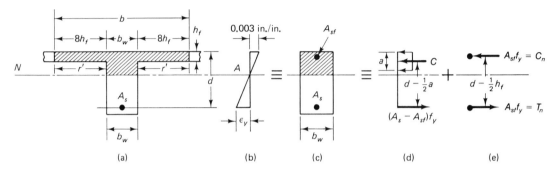

Figure 3.20 Stress and strain distribution in flanged sections design (T-beam transfer): (a) cross section; (b) strains; (c) transformed section; (d) part 1 forces; (e) part 2 forces.

Thus

$$C_n = 2r'h_f \times 0.85f_c' = 0.85f_c'(b - b_w)h_f,$$

where r' is the overhang length on each side of the web. The compressive force C_n is *equated* to a tensile force T_n for equilibrium such that $T_n = (A_{sf} \times f_y)$, where A_{sf} is an imaginary compressive steel area whose force capacity is equivalent to the force capacity of the compression flange overhang. Consequently, an equivalent area of A_{sf} of compression reinforcement to develop the overhang flanges would have a value

$$A_{sf} = \frac{0.85f_c'(b - b_w)h_f}{f_y} \tag{3.20}$$

For a beam to be considered a *real* T beam, the tension force $A_s f_y$ generated by the steel should be greater than the compression force capacity of the total flange area $0.85f_c'bh_f$. Thus

$$a = \frac{A_s f_y}{0.85f_c'b} > h_f \tag{3.21a}$$

or $$h_f < (1.18\overline{\omega}d = a) \tag{3.21b}$$

where $\overline{\omega} = (A_s/bd)(f_y f_c')$.

The concrete stress block is actually parabolic and extends to the neutral axis depth c. Consequently, from a theoretical point of view, if one were using a parabolic stress block, then Eq. 3.21b for a T beam can also be written

$$h_f < \frac{1.18\overline{\omega}d}{\beta_1} \tag{3.21c}$$

The percentage for the balanced condition in a T beam can be written

$$\rho_b = \frac{b_w}{b}(\bar{\rho}_b + \rho_f) \tag{3.22}$$

where $\bar{\rho}_b = \dfrac{0.85\beta_1 f_c'}{f_y} \dfrac{87{,}000}{87{,}000 + f_y}$

   ρ_f = reinforcement ratio for tension steel area necessary to develop the compressive strength of the overhanging flanges

$$\rho_f = 0.85 f_c'(b - b_w)\frac{h_f}{f_y b_w d}$$

As in the case of singly and doubly reinforced beams, the maximum allowable percentage of the steel ρ at the tension side should not exceed 75% of the balanced steel percentage ρ_b to ensure ductile failure. So in the case of a T beam,

$$\rho = \frac{A_s}{bd} \le 0.75\rho_b \tag{3.23}$$

A strain-incompatibility check is not needed, for the imaginary steel area A_{sf} is assumed to yield in all cases. To satisfy the requirement of minimum reinforcement so that the beam does not behave as nonreinforced,

$$\rho_w = \frac{A_s}{b_w d} \ge \frac{200}{f_y} \tag{3.24}$$

It should be noted that b_w is used in Eq. 3.24 instead of width b, which is used for singly or doubly reinforced beams.

As in the case of design and analysis of doubly reinforced sections, the reinforcement at the tension side is considered to be composed of two areas: A_{s1} to balance the rectangular block compressive force on area $b_w a$ and A_{s2} to balance the imaginary steel area A_{sf}. Consequently, the total nominal moment strength for parts 1 and 2 of the solution is

$$M_n = M_{n1} + M_{n2} \tag{3.25a}$$

$$M_{n1} = A_{s1}f_y\left(d - \frac{a}{2}\right) = (A_s - A_{sf})f_y\left(d - \frac{a}{2}\right) \tag{3.25b}$$

$$M_{n2} = A_{s2}f_y\left(d - \frac{h_f}{2}\right) = A_{sf}f_y\left(d - \frac{h_f}{2}\right) \tag{3.25c}$$

The design moment strength ϕM_n, which must be at least equal to the external factored moment M_u, becomes

$$M_u = \phi M_n = \phi\left[(A_s - A_{sf})f_y\left(d - \frac{a}{2}\right) + A_{sf}f_y\left(d - \frac{h_f}{2}\right)\right] \tag{3.26}$$

The flowchart (Fig. 3.21) presents the sequence of calculations for the analysis of the

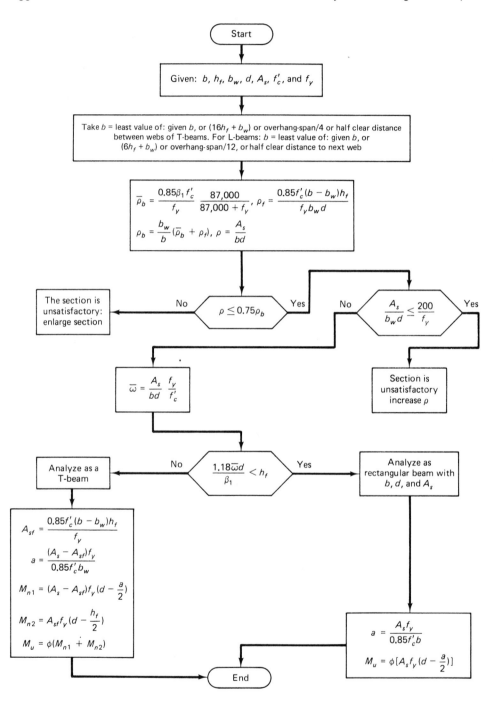

Figure 3.21   Flowchart for the analysis of T and L beams.

T beam. The following analysis example illustrates the nominal moment strength calculations for a typical precast T beam.

3.8.2 Example 3.7: Analysis of a T Beam for Moment Capacity

Calculate the nominal moment strength and the design ultimate moment of the precast T beam shown in Fig. 3.22 if the beam span is 30 ft (9.14 m). Given:

$f'_c = 4000$ psi (27.58 MPa), normalweight concrete
$f_y = 60,000$ psi (413.4 MPa)

Reinforcement area at the tension side:
(a) $A_s = 4.0$ in.^{2} (2580 mm^2)
(b) $A_s = 6.0$ in.^{2} (3870 mm^2)

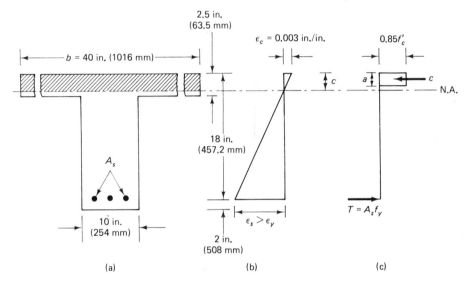

Figure 3.22 Geometry, strain, and force distributions in the T-beam of Ex. 3.7: (a) cross section; (b) strains; (c) stresses.

Solution

Flange width check is not necessary for a precast beam because the precast section can act independently, depending on the construction system.
Check for ρ_{max}:

$$\rho_{max} \le 0.75\rho_b$$

$$\bar{\rho}_b = \frac{0.85\beta_1 f'_c}{f_y} \frac{87,000}{87,000 + f_y} = \frac{0.85 \times 0.85 \times 4000}{60,000} \frac{87,000}{87,000 + 60,000} = 0.029$$

$$\rho_f = \frac{0.85f'_c(b - b_w)h_f}{f_y b_w d}$$

$$= \frac{0.85 \times 4000(40 - 10) \times 2.5}{60,000 \times 10 \times 18} = 0.024$$

$$\rho_b = \frac{b_w}{b}(\bar{\rho}_b + \rho_f) = \frac{10}{40}(0.029 + 0.024) = 0.013$$

$$\rho_{max} = 0.75\rho_b = 0.010$$

$$\rho_{min} = \frac{200}{60,000} = 0.0033$$

(a) $A_s = 4$ in.^{2}

$$\rho_w = \frac{A_s}{b_w d} = \frac{4.0}{10 \times 18} = 0.0222 > \rho_{min} = 0.0033 \qquad \text{O.K.}$$

$$\rho = \frac{A_s}{bd} = \frac{4.0}{40 \times 18} = 0.0006 < \rho_{max} \qquad \text{O.K.}$$

Check whether the section will act as a T beam.

$$\bar{\omega} = \frac{A_s f_y}{bd f'_c} = \frac{4.0}{40 \times 18}\left(\frac{60}{4}\right) = 0.083$$

$$c = \frac{1.18\bar{\omega}d}{\beta_1} = \frac{1.18 \times 0.083 \times 18}{0.85} = 2.10 \text{ in.} < (h_f = 2.5 \text{ in.})$$

Therefore the beam can be analyzed as a rectangular beam, using b, d, and A_s.

$$M_n = A_s f_y\left(d - \frac{a}{2}\right)$$

$$a = \frac{4.0 \times 60,000}{0.85 \times 4000 \times 40} = 1.765 \text{ in.}$$

$$M_n = 4.0 \times 60,000\left(18 - \frac{1.765}{2}\right) = 4,108,200 \text{ in.-lb}$$

$$M_u = \phi M_n = 0.9 \times 4,108,200 \text{ in.} = 3,697,380 \text{ in.-lb} (417.8 \text{ kN-m})$$

(b) $A_s = 6.0$ in.^{2}.

$$\rho_w = \frac{A_s}{b_w d} = \frac{6.0}{10 \times 18} = 0.033 > \rho_{min} = 0.0033 \qquad \text{O.K.}$$

$$\rho = \frac{A_s}{bd} = \frac{6.0}{40 \times 18} = 0.0083 < \rho_{max} = 0.010 \qquad \text{O.K.}$$

$$\bar{\omega} = \frac{6.0}{40 \times 18}\left(\frac{60}{4}\right) = 0.125$$

$$\frac{1.18\bar{\omega}d}{\beta_1} = \frac{1.18 \times 0.125 \times 18}{0.85} = 3.124 > (h_f = 2.5)$$

Thus the neutral axis is *below* the flange. The beam must be treated as a T beam or equivalent doubly reinforced rectangular beam with imaginary compression steel area A_{sf}.

$$A_{sf} = \frac{0.85f_c'(b - b_w)h_f}{f_y}$$

$$= \frac{0.85 \times 4000(40 - 10) \times 2.5}{60,000} = 4.25 \text{ in.}^2$$

$$a = \frac{(A_s - A_{sf})f_y}{0.85f_c'b_w}$$

$$a = \frac{(6.00 - 4.25) \times 60,000}{0.85 \times 4000 \times 10}$$

$$= 3.09 \text{ in.}$$

$$M_{n1} = (A_s - A_{sf})f_y\left(d - \frac{a}{2}\right)$$

$$= (6.00 - 4.25) \times 60,000\left(18 - \frac{3.09}{2}\right)$$

$$= 1,727,780 \text{ in.-lb}$$

$$M_{n2} = A_{sf}f_y\left(d - \frac{h_f}{2}\right)$$

$$= 4.25 \times 60,000\left(18 - \frac{2.5}{2}\right) = 4,271,250 \text{ in.-lb}$$

$$M_n = M_{n1} + M_{n2} = (1,727,780 + 4,271,150) = 6,000,000 \text{ in.-lb}$$

$$= 6000.00 \text{ in.-kips}$$

$$M_u = \phi M_n = 0.9 \times 6,000,000 = 5,400,000 \text{ in.-lb} \ (610.2 \text{ kN-m})$$

3.8.3 TRIAL-AND-ADJUSTMENT PROCEDURE FOR THE DESIGN OF FLANGED SECTIONS

The slab thickness h_f of the flange overhang is known at the outset, for the slab is designed first. Also available is the external factored moment M_u at midspan. The trial-and-adjustment steps for proportioning the web of the beam section can be summarized as follows.

1. Choose a singly reinforced beam section that can resist the external factored moment M_u and the moment due to self-weight. Remember that a T section or an L section would have a smaller size or depth than a singly reinforced section.

2. Check whether the span/depth ratio is reasonable—that is, between 12 and 18. If not, adjust the preliminary section.
3. Calculate the flange width on the basis of the criteria in Section 3.8.
4. Determine if the neutral axis is within or outside the flange, where the neutral axis depth $c = 1.18\overline{\omega}d/\beta_1$ for rectangular singly reinforced sections.
   (a) If $c < h_f$, the beam should be treated as a singly reinforced beam with a width b equivalent to the flange width determined in step 2.
   (b) If $c > h_f$ and the equivalent block depth a is also $> h_f$, design as a T beam or an L beam, as the case may be.
5. Find the equivalent compressive steel area A_{sf} for the flange overhang and analyze the assumed section as in Ex. 3.7(b). Calculate the nominal resisting capacities M_{n1} and M_{n2}.
6. Repeat steps 4 and 5 until the calculated $\phi M_n = \phi(M_{n1} + M_{n2})$ is close in value to the factored moment M_u and verify that the assumed self-weight of the web is correct.
7. *Alternatively,* the first trial section can be chosen using a moment factor $R'' > R$ in step 3(a) of Section 3.5 for singly reinforced beams such that $R'' \simeq 1.35 - 1.50R$. Select a trial section depth from $M_n = R''bd^2$ and proceed to analyze the section.

3.8.4 Example 3.8: Design of an End-Span L Beam

A roof-garden floor is composed of a monolithic one-way slab system on beams as in Fig. 3.23. The effective beam span is 35 ft (10.67 m) and all beams are spaced at 7 ft 6 in. (2.29 m) center to center. The floor supports a 6 ft 4 in. (1.52 m) depth of soil in addition to its self-weight. Assume also that the slab edges support a 12-in.-wide, 7-ft wall weighing 840 lb per linear foot. Design the midspan section of the edge spandrel L beam AB, assuming that the moist soil weighs 125 lb/ft³ (2.56 tons/m³). Given:

$$f'_c = 3000 \text{ psi (20.68 MPa), normalweight concrete}$$
$$f_y = 60,000 \text{ psi (413.7 MPa)}$$

Solution

  Slab design

The weight of soil = $6.33 \times 125 = 791$, say 800 psf (38.32 kPa). Assume a slab thickness $h = 4$ in. (101.6 mm) = $4/12 \times 150 = 50$ psf. $d = h - (\frac{3}{4}$ in. cover $+ \frac{1}{2}$ diameter of No. 4 bars) = $4.0 - 1.0 = 3.0$ in.

Factored load intensity $w_u = 1.4(800 + 50) = 1190$ lb/ft² (57.0 kPa)

From the ACI code the negative moment for the first interior support of a continuous slab is

$$-M_u = \frac{w_u l_n^2}{12} = \frac{1190(7.5)^2}{12} \times 12 = 66,938 \text{ in.-lb}$$

The required slab negative moment strength

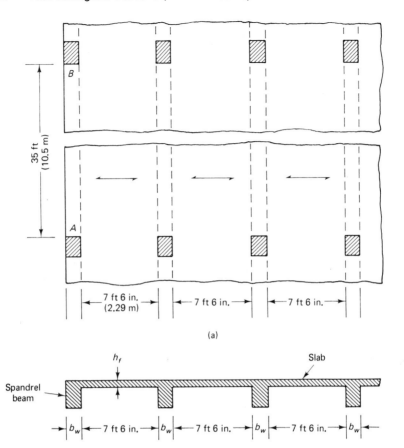

Figure 3.23 Spandrel beam *AB* design in Ex. 3.8: (a) floor plan; (b) transverse section.

$$-M_n = \frac{66,938}{0.9} = 74,376 \text{ in.-lb}$$

$$M_n = A_s f_y \left(d - \frac{a}{2} \right)$$

Assume that $(d - a/2) \simeq 0.9d = 0.9 \times 3.0 = 2.7$ in.

$$A_s = \frac{74,376}{60,000 \times 2.7} = 0.459 \text{ in.}^2 \text{ on a 12-in. strip}$$

Try $b = 12$ in., $d = 3$ in., and No. 5 bars at 7.5 in. center to center. $A_s = 0.496$ in.^{2}

$$a = \frac{A_s f_y}{0.85 f_c' b} = \frac{0.496 \times 60,000}{0.85 \times 3,000 \times 12} = 0.97 \text{ in.}$$

$$\text{Nominal resisting moment } M_n = 0.496 \times 60{,}000\left(3.0 - \frac{0.97}{2}\right) = 74{,}846 \text{ in.-lb}$$

$$> \text{required } M_n = 74{,}376 \text{ in.-lb} \qquad \text{O.K.}$$

$$\rho = \frac{0.496}{12 \times 3} = 0.0138$$

$$\bar{\rho}_b = \frac{0.85\beta_1 f_c'}{f_y}\left(\frac{87{,}000}{87{,}000 + f_y}\right) = \frac{0.85 \times 0.85 \times 3000}{60{,}000}\left(\frac{87{,}000}{87{,}000 + f_y}\right) = 0.0214$$

$$\text{Maximum allowable } \rho = 0.75\rho_b = 0.016 > 0.0138 \qquad \text{O.K.}$$

Similarly, for the positive moment, $+M_u = w_u l_n{}^2/16$, requiring No. 5 bars at 10 in. center to center. Use No. 5 bars at $7\frac{1}{2}$ in. center to center main negative reinforcement (15.9-mm diameter at 190.5-mm spacing) and No. 5 bars at 10 in. center to center for main positive reinforcement.

$$\text{Temperature steel} = 0.0018bh = 0.0018 \times 12 \times 4.0 = 0.0864 \text{ in.}^2$$

$$\text{Maximum allowable spacing} = 5h = 5 \times 4 = 20 \text{ in. (508.0 mm)}$$

Use No. 3 bars at 15 in. $= 0.088$ in.^{2} (9.53 mm diameter at 381 mm spacing) for temperature.

Beam web design

In order to choose a trial web section, assume that $d = l_n/18$ for deflection or

$$d = \frac{35.0 \times 12}{18} = 23.33 \text{ in.}$$

Assume that $h = 26$ in. (660.4 mm), $d = 22.5$ in. (571.6 mm), and $b_w = 14$ in. (355.6 mm).

$$\text{Load area on L beam } AB = \frac{7.5}{2} + \frac{14}{12} = 4.92 \text{ ft}$$

$$\text{Superimposed working } w_w = (4.92 - 1.0) \times 800 = 3136 \text{ lb/ft}$$

$$\text{Slab weight} = \frac{4.0}{12} \times 150 \times 4.92 = 246 \text{ lb/ft}$$

$$\text{Weight of beam web} = \frac{14(26 - 4)}{144} \times 150 = 321 \text{ lb/ft}$$

$$\text{7-ft wall weight} = 840 \text{ lb/ft}$$

$$\text{Total service load} = 3136 + 246 + 321 + 840 = 4543 \text{ lb/ft}$$

$$\text{Factored load } w_u = 1.4 \times 4543 = 6360 \text{ lb/ft}$$

$$\text{Factored external moment } M_u = \frac{w_u l_n{}^2}{11} = \frac{6360(35.0)^2}{11} \times 12 = 8{,}499{,}273 \text{ in.-lb}$$

$$M_n = \frac{M_u}{\phi} = \frac{8{,}499{,}273}{0.9} = 9{,}443{,}637 \text{ in.-lb (1067.06 kN-m)}$$

To determine whether the beam is an actual L beam or not, it is necessary to find if the neutral axis falls outside the flange. Consequently, the area of the tension steel A_s at midspan has to be assumed. If a rectangular section is initially assumed with an appropriate moment arm $jd \simeq 0.85d = 0.85 \times 22.5 = 19.3$ in.,

$$M_n = A_s f_y jd \quad or \quad 9{,}443{,}637 = A_s \times 60{,}000 \times 19.3$$

$$A_s = \frac{9{,}443{,}637}{60{,}000 \times 19.3} = 8.16 \text{ in.}^2$$

Assume eight No. 9 bars in two layers = 8.0 in.^{2} (5160 mm^2).

$$\rho = \frac{A_s}{bd} \quad \text{where } b = b_w + 6h_f = 14 + 6 \times 4.0 = 38 \text{ in. (965.2 mm)}$$

$$\rho = \frac{8.0}{38 \times 22.5} = 0.00936$$

$$\overline{\omega} = \rho \frac{f_y}{f_c'} = 0.00936 \times \frac{60{,}000}{3000} = 0.187$$

Depth of neutral axis $c = \dfrac{1.18\overline{\omega}d}{\beta_1} = \dfrac{1.18 \times 0.187 \times 22.5}{0.85}$

or

$$c = 5.84 \text{ in.} > 4.0 \text{ in.}$$

$$a = \beta_1 c = 0.85 \times 5.84 = 4.96 > 4.0 \text{ in.}$$

So the section is an L beam because the neutral axis is below the flange, as shown in Fig. 3.24.

For $f_c' = 3000$ psi and $f_y = 60{,}000$ psi, $\overline{\rho}_b = 0.0214$ and

$$A_{sf} = \frac{h_f(b - b_w)0.85f_c'}{f_y} = \frac{4.0(38 - 14) \times 0.85 \times 3000}{60{,}000}$$

$$= 4.08 \text{ in.}^2 \text{ (2631.6 mm}^2\text{)}$$

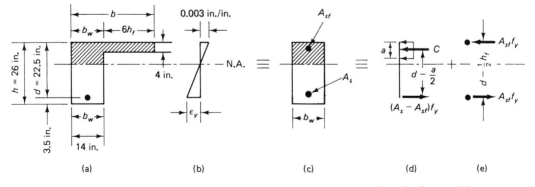

Figure 3.24 Forces and stresses in L-beams: (a) cross section; (b) strain diagram; (c) transformed section; (d) part 1 forces; (e) part 2 forces.

$$\rho_f = \frac{A_{sf}}{b_w d} = \frac{4.08}{14 \times 22} = 0.01295$$

$$\rho_b = (\bar{\rho}_b + \rho_f)\frac{b_w}{b} = (0.0214 + 0.01295)\frac{14}{38} = 0.01266$$

$0.75\rho_b = 0.00949 >$ actual $\rho = 0.00936$

Thus the section is underreinforced and satisfies the ACI code requirements.

$$a = \frac{(A_s - A_{sf})f_y}{0.85f_c'b_w} = \frac{(8.0 - 4.08)60,000}{0.85 \times 3000 \times 14} = 6.59 \text{ in. (167.4 mm)}$$

$$M_n = (A_s - A_{sf})f_y\left(d - \frac{a}{2}\right) + A_{sf}f_y\left(d - \frac{1}{2}h_f\right)$$

$$= (8.0 - 4.08)60,000\left(22.5 - \frac{6.59}{2}\right) + 4.08 \times 60,000\left(22.5 - \frac{4.0}{2}\right)$$

$$= 9,535,416 \text{ in.-lb}$$

Design moment $M_u = 0.9 \times 9,535,416$ in.-lb $= 8,581,874$ in.-lb

Actual factored $M_u = 8,499,273$ in.-lb $< 8,581,874$ in.-lb

Adopt the design. Flexural reinforcement details for the L-beam AB are shown in Fig. 3.25.

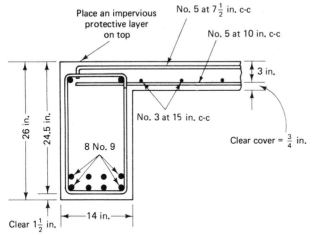

Place an impervious protective layer on top

No. 5 at $7\frac{1}{2}$ in. c-c

No. 5 at 10 in. c-c

3 in.

No. 3 at 15 in. c-c

Clear cover = $\frac{3}{4}$ in.

26 in.

24.5 in.

8 No. 9

14 in.

Clear $1\frac{1}{2}$ in.

Figure 3.25 Midspan section flexural reinforcement details for beam AB of Ex. 3.8.

3.9 TWO-WAY SLAB AND PLATE FLOOR SYSTEMS

Supported floor systems are usually constructed of reinforced concrete cast in place. Two-way slabs and plates are those panels in which the dimensional ratio of length to width is less than 2. The analysis and design of framed floor slab systems represented in Fig. 3.26 encompasses more than one aspect. The present state of knowledge

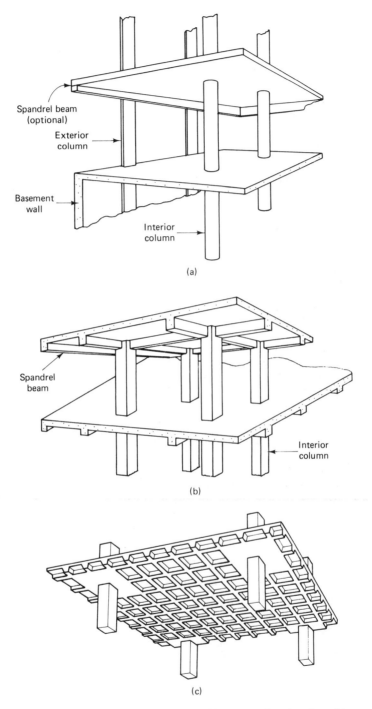

Figure 3.26 Two-way-action floor systems: (a) two-way flat-plate floor; (b) two-way slab floor on beams; (c) waffle slab floor.

permits reasonable evaluation of (a) the moment capacity, (b) the slab–column shear capacity, and (c) serviceability behavior as determined by deflection control and crack control. It is to be noted that flat plates are slabs supported directly on columns without beams, as shown in Fig. 3.26a compared to Fig. 3.26b for slabs on beams, or Fig. 3.26 for waffle slab floors.

The scope of this book does not permit coverage of two-way slab and plate design systems. Several methods are available for such a design, such as the ACI direct design method, the ACI equivalent frame method, the limit state at failure methods, including the yield line theory, the strip method, and others. The ACI methods are based on redistribution of moments from the higher stressed regions at the supports to the lower stressed regions at midspan after transformation of the slab or the plate into part of the horizontal element of an equivalent frame strip in each of the two perpendicular spans. Extensive coverage of the various methods for the analysis and design of two-way floor slabs and plates is given in the author's book in Ref. 3.3 as well as in other advanced texts.

SELECTED REFERENCES

3.1   Hognestad, E. N., N. W. Hanson, and D. McHenry, "Concrete Stress Distribution in Ultimate Strength Design," *Journal of the American Concrete Institute,* Proc. Vol. 52, December 1955, pp. 455–479.

3.2   Mattock, A. H., L. B. Kriz, and E. N. Hognestad, "Rectangular Stress Distribution in Ultimate Strength Design," *Journal of the American Concrete Institute,* Proc. Vol. 58, February 1961, pp. 825–928.

3.3   Nawy, E. G., "Reinforced Concrete—A Fundamental Approach," Prentice-Hall, Englewood Cliffs, NJ, 1985, 720 pp.

3.4   Nawy, E. G., "Strength, Serviceability and Ductility," Chapter 12 in *Handbook of Structural Concrete,* Pitman Books, London/McGraw-Hill, New York, 1983, 1968 pp.

3.5   Whitney, C. S. and E. Cohn, "Guide for Ultimate Strength Design of Reinforced Concrete," *Journal of the American Concrete Institute,* Proc. Vol. 53, November 1956, pp. 455–475.

3.6   Whitney, C. S., "Plastic Theory of Reinforced Concrete Design," *Transactions of the ASCE,* Vol. 107, 1942, pp. 251–326.

3.7   ACI Committee 318, "Building Code Requirements for Reinforced Concrete, ACI Standard 318–83," American Concrete Institute, Detroit, 1983, 111 pp.

3.8   "Commentary on Building Code Requirements for Reinforced Concrete," American Concrete Institute, 1983, 155 pp.

3.9   ACI Committee 340, *Design Handbook,* Vol. 1, Special Publication No. 17, American Concrete Institute, Detroit, 1984, 374 pp.

PROBLEMS FOR SOLUTION

3.1   Calculate the nominal moment strength of the beam sections shown in Fig. 3.27. Given:

$$f_c' = 3000 \text{ psi } (20.68 \text{ MPa})$$

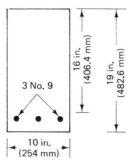

Figure 3.27

$f'_c = 6000$ psi (41.36 MPa)

$f'_c = 9000$ psi (62.10 MPa)

$f_y = 60,000$ psi (413.7 MPa)

3.2 Calculate the safe load that the beam in Fig. 3.28 can carry. Given:

$f'_c = 4000$ psi (27.58 MPa), normalweight concrete

$f_y = 60,000$ psi (413.7 MPa)

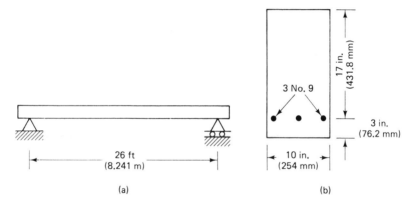

(a)   (b)

Figure 3.28

3.3 Design a one-way slab to carry a live load of 100 psf and an external dead load of 50 psf. The slab is simply supported over a span of 12 ft. Given:

$f'_c = 4000$ psi (27.58 MPa), normalweight concrete

$f_y = 60,000$ psi (413.7 MPa)

3.4 Calculate the stresses in the compression steel, f'_s, for the cross sections shown in Fig. 3.29. Given:

$f'_c = 6000$ psi (41.37 MPa), normalweight concrete

$f_y = 60,000$ psi (413.7 MPa)

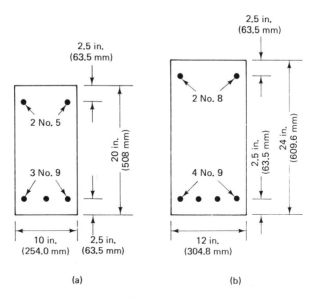

Figure 3.29

3.5 Calculate the nominal moment strength of the beam sections of Problem 3.2. Assume two No. 6 bars for compression reinforcement.

3.6 At failure, determine whether the precast sections shown in Fig. 3.30 will act similarly to rectangular sections or as flanged sections. Given:

f'_c = 4000 psi (27.58 MPa), normalweight concrete

f_y = 60,000 psi (413.7 MPa)

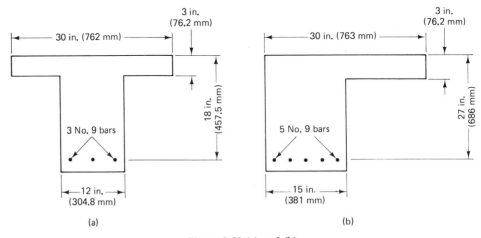

Figure 3.30 (a) and (b)

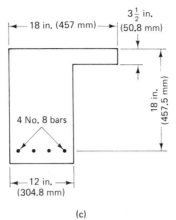

(c)   **Figure 3.30 (c)**

3.7 Find the nominal moment strength M_n of the flanged beams shown in Fig. 3.30. Given:

$f'_c = 4000$ psi (27.58 MPa)

$f_y = 60,000$ psi (413.7 MPa)

4

Development Length
in Steel Reinforcement

4.1 BOND STRESS DEVELOPMENT

The change in stress along the length of a bar in a beam due to the variation of moment along the span is shown schematically in Fig. 4.1b. If jd is the lever arm of the couple T due to moment M, then $T = M/jd$. In terms of the moment difference between cracked sections 1 and 2,

$$dT = \frac{dM}{jd} \qquad (4.1a)$$

Also,

$$dT = \mu d \times \Sigma o \qquad (4.1b)$$

where Σo is the total circumference of all the reinforcement subjected to the bond stress pull, to get $dM/dx = \mu \Sigma ojd$; because $dM/dx = $ shear V,

$$\mu = \frac{V}{\Sigma ojd} \qquad (4.1c)$$

Equation 4.1c is primarily of academic importance, for it is only indirectly accounted for in the development length l_d shown in Fig. 4.1. The development length l_d as a function of the size and yield strength of the reinforcement determines the resistance of the bars to slippage—hence the magnitude of the beam's failure capacity. It has been verified by tests that the bond strength μ is a function of the compressive strength of concrete such that

$$\mu = k\sqrt{f_c'} \qquad (4.2a)$$

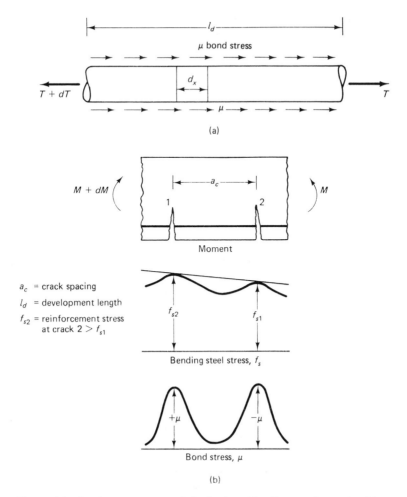

Figure 4.1   Bond stress across a reinforcing bar: (a) pull-out anchorage bond in a bar; (b) flexural bond.

where k is a constant. If the bond strength equals or exceeds the yield strength of a bar of cross-sectional area $A_b = \pi d_b^2/4$, then

$$\pi d_b l_d \mu \geq A_b f_y \tag{4.2b}$$

From Eqs. 4.2a and 4.2b

$$l_{db} = k_1 \frac{A_b f_y}{\sqrt{f_c'}} \tag{4.3}$$

where k_1 is a function of the geometrical property of the reinforcing element and the relationship between bond strength and compressive strength of concrete.

    Equation 4.3 is consequently the basic model for defining the minimum development length of bars in structural elements. The ACI code specifies k_1 values for

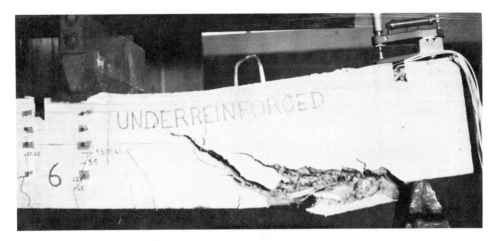

Photo 34 Bond failure at support of simply supported beam. (Tests by Nawy, et al.)

Photo 35 Bond failure and destruction of concrete cover at rupture load. (Test by Nawy, et al.)

various bar sizes and bond stresses in tension as well as in compression. These values are the result of extensive experimental verifications, particularly those reported by Ferguson et al. in Ref. 4.1.

4.1.1 Development of Deformed Bars in Tension

A reinforcing bar should have a sufficient embedment length l_d to prevent bond failure. The factor k_1 in Eq. 4.3 has different values for different bar sizes. The ACI code includes modifying multipliers for top bars because such bars have a lesser confining cover effect, hence lesser bond strength, than bottom bars. Modifying multipliers are also provided for lightweight concrete and for bars of yield strength higher than 60,000 psi. The basic development length should be as follows but not less than 12 in. (304.8 mm).

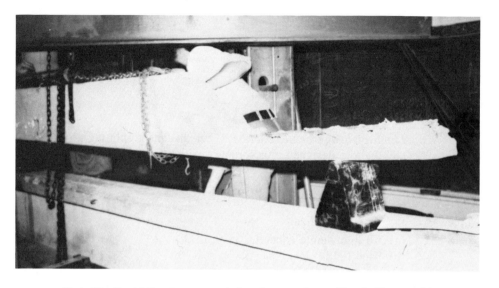

Photo 36   Bond failure in an over-reinforced concrete beam. (Tests by Nawy, et al.)

No. 11 bars and smaller:

$$l_{db} = 0.04 \frac{A_b f_y}{\sqrt{f_c'}} \qquad (4.4a)$$

and
$$l_{db} \geq 0.0004 d_b f_y \qquad (4.4b)$$

where *db* is the bar diameter.

No. 14 bar:

$$l_{db} = 0.085 \frac{f_y}{\sqrt{f_c'}} \qquad (4.4c)$$

No. 18 bar:

$$l_{db} = 0.110 \frac{f_y}{\sqrt{f_c'}} \qquad (4.4d)$$

Deformed wire (but not including wire fabric):

$$l_{db} = 0.03 \frac{d_b f_y}{\sqrt{f_c'}} \qquad (4.4e)$$

4.1.2 Modifying Multipliers of Development Length l_d for Bars in Tension

The basic development length l_{db} must be multiplied by the applicable λ_d factor for the particular conditions to achieve the necessary embedment length, so that development length $l_d = l_{db}\lambda_d$.

1. *Top reinforcement:*

$$\lambda_d = 1.4$$

2. *Reinforcement with $f_y > 60,000$ psi:*

$$\lambda_d = 2 - \frac{60,000}{f_y}$$

3. *Lightweight aggregate concrete:* When the tensile splitting strength f_{ct} is specified and the mix designed in accordance with ACI:

$$\lambda_d = \frac{6.7\sqrt{f_c'}}{f_{ct}} \geq 1.0$$

If f_{ct} is not specified:
(a) All lightweight concrete: $\lambda_d = 1.33$
(b) Sand-lightweight concrete: $\lambda_d = 1.18$

    The basic development length l_d can be reduced by using a reduction multiplier for certain conditions as follows.

1. *Reinforcement spaced laterally:* At least 6 in. on center and 3 in. minimum from face of member to edge bar: $\lambda_d = 0.8$.
2. *Required A_s < provided A_s:*

$$\lambda_d = \frac{\text{required } A_s}{\text{provided } A_s}$$

3. *Confined reinforcement:* confinement within spiral reinforcement not less than $\frac{1}{4}$ in. diameter and not more than 4 in. of pitch: $\lambda_d = 0.75$.

4.1.3 Development of Deformed Bars in Compression and the Modifying Multipliers

Bars in compression require shorter development length than bars in tension. This situation is due to the absence of the weakening effect of the tensile cracks. Thus the expression for the basic development length is

$$l_{db} = 0.02 \frac{d_b f_y}{\sqrt{f_c'}} \tag{4.5a}$$

and

$$l_{db} \geq 0.0003 d_b f_y \tag{4.5b}$$

with the modifying multiplier for

1. Excess reinforcement: $\lambda_d = \text{required } A_s/\text{provided } A_s$
2. Spirally enclosed reinforcement: $\lambda_d = 0.75$

If bundled bars are used in tension or compression, l_d must be increased by 20% for three-bar bundles and 33% for four-bar bundles.

4.1.4 Example 4.1: Embedment Length of Deformed Bars

Calculate the required embedment length of the deformed bars of the following three cases.

(a) No. 7 bar (22.2 mm diameter) top reinforcement:
   $f_y = 60,000$ psi (413.7 MPa)
   $f'_c = 4500$ psi (31.03 MPa)

The section was overdesigned by providing a steel area $A_s = 7.2$ in.^{2} (4644 mm^2) instead of the required $A_s = 6.9$ in.^{2}

(b) $\frac{1}{8}$-in.-diameter deformed wire in tension:
   $f_y = 80,000$ psi (551.6 MPa)
   $f'_c = 5000$ psi (34.47 MPa)
(c) No. 8 bar (28.6 mm diameter) in compression:
   $f_y = 60,000$ psi (413.7 MPa)
   $f'_c = 4000$ psi (27.6 MPa)

Solution

$$\text{(a) } l_{db} = \frac{0.04A_b f_y}{\sqrt{f'_c}} = \frac{0.04 \times 0.6 \times 60,000}{\sqrt{4500}} = 21.47 \text{ in.}$$

or

$$l_{db} = 0.0004d_b f_y = 0.0004 \times 0.875 \times 60,000 = 21.0 \text{ in.}$$

Therefore

Controlling basic development length $= 21.47$ in.

Modifying multiplier for top reinforcement $\lambda_d = 1.4$

Modifier for overdesign $\lambda_d = \dfrac{6.9}{7.2} = 0.958$

And so,

Minimum embedment
or development length $l_{db} = 1.4 \times 0.958 \times 21.47$

$$= 28.8 \text{ in. (732 mm)} \qquad \text{say 30 in.} > 12 \text{ in.} \qquad \text{O.K.}$$

$$\text{(b) } l_{db} = \frac{0.03d_b f_y}{\sqrt{f'_c}} = \frac{0.03 \times 0.125 \times 80,000}{\sqrt{5000}} = 4.24 \text{ in.}$$

For $f_y > 60,000$ psi,

$$\lambda_d = 2 - \frac{60,000}{80,000} = 1.25$$

Minimum development length $l_d = 1.25 \times 4.24 = 5.30$ in. < 12.0 in.

Use $l_d = 12$ in. (305 mm).

$$\text{(c)} \quad l_{db} = \frac{0.02 d_b f_y}{\sqrt{f_c'}} = \frac{0.02 \times 1.0 \times 60{,}000}{\sqrt{4500}} = 17.89 \text{ in.}$$

or

$$l_{db} = 0.0003 d_b f_y = 0.0003 \times 1.0 \times 60{,}000 = 18.0 \text{ in.}$$

Therefore

Minimum development length $l_d = 18.0$ in. (457 mm)

4.1.5 Mechanical Anchorage and Hooks

Hooks are used when space limitation in a concrete section does not permit the necessary straight embedment length. Hooks in structural members are placed relatively close to the free surface of a concrete element, where splitting forces proportional to the total bar force may determine the hook capacity. The standard hook does *not* develop the tension yield strength of the bar. If l_{hb} is the basic development length for the standard hook in tension, additional embedment length must be incorporated to give a total length l_{dh} not less than $8d_b$ or 6 in., whichever is greater. l_{dh} length is shown in Fig. 4.2. The length l_{hb} varies with the bar size, reinforcement yield strength, and compressive strength of concrete. For $f_y = 60{,}000$-psi steel,

$$l_{hb} = 1200 \, \frac{d_b}{\sqrt{f_c'}} \tag{4.6}$$

where d_b is the diameter of the hook bar.

Modifying Multipliers for Hooks in Tension

1. *Yield strength f_y effect:* For a yield different than 60,000, $\lambda_d = f_y/60{,}000$.
2. *Concrete cover effect:* For No. 11 bars and smaller, side cover normal to the plane of hook not less than $2\frac{1}{2}$ in. and for 90° hook with cover on bar extension beyond the hook not less than 2 in., $\lambda_d = 0.7$ (see Fig. 4.2c).
3. *Ties or stirrups:* For No. 11 bars and smaller, hook enclosed vertically or horizontally within stirrups or ties spaced not greater than $3d_b$, where d_b is diameter of hook bar, $\lambda_d = 0.8$.
4. *Excess reinforcement:* Where anchorage or development for f_y is not specifically required but the reinforcement area A_s used in excess of A_s required for analysis,

$$\lambda_d = \frac{\text{required } A_s}{\text{provided } A_s}$$

5. *Bars developed by standard hooks at discontinuous ends:* If the concrete cover is less than $2\frac{1}{2}$ in., bars *should* be *enclosed* within ties or stirrups along the full

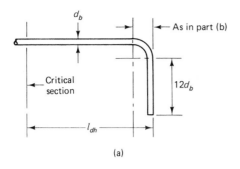

(a)

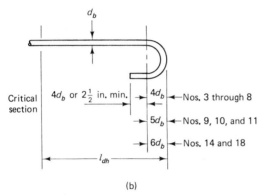

(b)

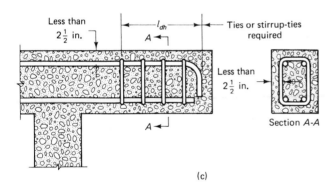

(c)

Figure 4.2 Standard bar hook details: (a) 90° hook; (b) 180° hook; (c) hook in small concrete cover.

development length l_{dh} spaced at no greater than $3d_b$; for this case, $\lambda_d = 0.8$ from item 3; modifying multiplier shall not apply.

6. *Lightweight concrete: $\lambda_d = 1.3$.* It should be noted that hooks cannot be considered effective in developing bars in compression. The total development or embedment length

$$l_{dh} = l_{hb} \times \lambda_d \qquad (4.7)$$

Figure 4.2a and b shows details of standard 90° and 180° hooks used in axial tension or bending tension and Fig. 4.2c gives details of bar hooks susceptible to concrete splitting when the cover is small—that is, less than $2\frac{1}{2}$ in. Confinement is enhanced through the use of *closed* ties or stirrups. No distinction is made between a top bar and a bottom bar if hooks are used.

4.1.6 Example 4.2: Embedment Length for a Standard 90° Hook

Compute the development length required for the top bars of a lightweight concrete beam extending into the column support shown in Fig. 4.3, assuming No. 9 reinforcing bars (28.6 mm diameter) hooked at the end. The concrete cover is 2 in (508 mm). Given:

$$f_c' = 5000 \text{ psi } (34.47 \text{ MPa})$$

$$f_y = 60{,}000 \text{ psi } (413.7 \text{ MPa})$$

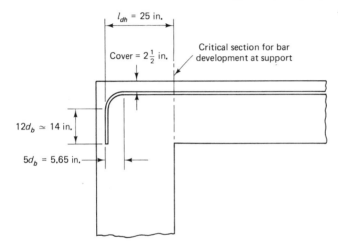

Figure 4.3 Hook embedment detail.

Solution

Top bars for hooks behave similarly to bottom bars; so no modifier is needed. For No. 9 bars, $d_b = 1.128$ in. (28.65 mm).

$$\text{Basic development length } l_{hb} = 1200 \frac{d_b}{\sqrt{f_c'}}$$

or

$$l_{hb} = \frac{1200 \times 1.128}{\sqrt{5000}} = 19.14 \text{ in.}$$

For lightweight concrete, $\lambda_d = 1.3$.

$$l_{dh} = 1.3 \times 19.14 = 24.88 \text{ in.} > 8d_b \text{ or 6 in.} \qquad \text{O.K.}$$

Use a 90° hook with embedment length $l_{dh} = 25$ in. (635 mm) beyond the critical section (face of support). Figure 4.3 shows the geometrical details of the hook.

4.2 DEVELOPMENT OF FLEXURAL REINFORCEMENT IN CONTINUOUS BEAMS

As noted, reinforcing bars should be adequately embedded in order to prevent serious bar slippage, resulting in bond pull-out failure. The critical locations for bar discontinuance are points along the structural member where there is a rapid drop in the bending moment or stress, such as the inflection points in a bending moment diagram of a continuous beam.

Tension reinforcement can be developed by bending the lower tension bars at a 45° inclination across the web of the beam and the bars should be anchored or made continuous with the reinforcing bars on the top of the member. To ensure full development, reinforcement has to be extended beyond the point at which it is no longer required to resist flexure for a distance equal to the effective depth d or 12 d_b, whichever is greater, except for supports of simple span beams or at the free end of a cantilever. Figure 4.4 shows details of flexural reinforcement development in typical continuous beams for both positive and negative steel reinforcement.

Here are general guidelines for full development of the reinforcement and for ensuring continuity in the case of continuous beams.

1. At least *one-third* of the positive moment reinforcement in simple beams and *one-fourth* of the positive moment reinforcement in continuous beams should be extended at least 6 in. into the support without being bent.
2. At simple supports as in Fig. 4.5a and at points of inflection as in Fig. 4.5b, the positive moment reinforcement should be limited to such a diameter that the development length

$$l_d \leq \frac{M_n}{V_u} + l_a \qquad (4.8)$$

where M_n = nominal moment strength where all the reinforcement is stressed to f_y

V_u = factored shear force at the section under consideration

and l_a equals effective depth d or $12d_b$, whichever is greater, d_b being the bar diameter.

Equation 4.8 imposes a design limitation on the flexural bond stress in areas of large shear and small moment in order to prevent splitting. Such a condition exists in short-span heavily loaded simple beams. Thus the bar diameter for positive moment is so chosen that even if length AC to the critical section in Fig. 4.5a is larger than length AB, the bar size must be limited such that $l_d \leq 1.3 (M_n/V_u + l_a)$. For confining reactions, such as at simple supports, the value M_n/V_u in Eq. 4.8 is increased by 30%.

3. At least *one-third* of the total tension reinforcement provided for negative bending moment at the support should be extended beyond the inflection point not less

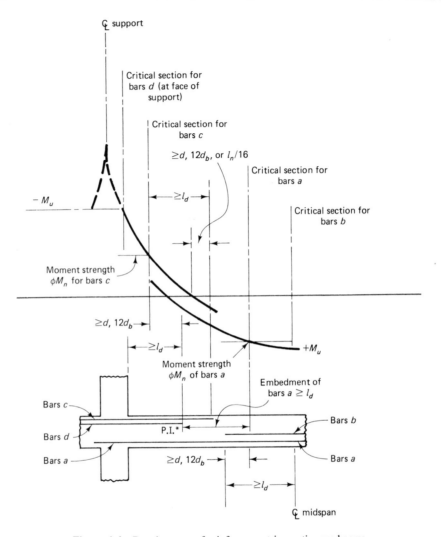

Figure 4.4 Development of reinforcement in continuous beams.

than the effective depth d of the member, $12d_b$, or $\frac{1}{16}$ of the clear span, whichever has the largest value.

4. Web stirrups should be carried as close to the compression and tension surfaces of the member as the minimum concrete cover requirements allow. The ends of the stirrups without hooks should have an embedment of at least $d/2$ above or below the compression side of the member for full development length l_d but not less than 12 in. or $24d_b$. For stirrups with hook ends, the total embedment length should equal $0.5l_d$ plus the standard hook.

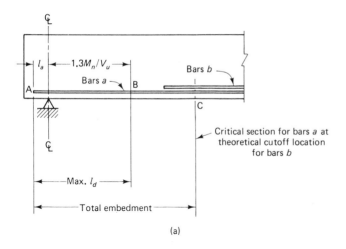

(a)

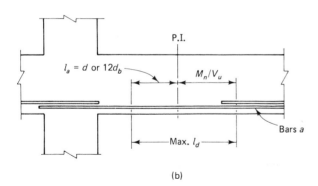

(b)

Figure 4.5   Cutoff points for reinforcement: (a) simply supported beams; (b) continuous beams.

A typical detail of cutoff points for continuous one-way beam and joist construction from Ref. 4.2 is given in Fig. 4.6. Typical cutoff points for one-way slabs are shown in Fig. 4.7 and for beams with diagonal tension stirrups in Fig. 4.8.

4.3 SPLICING OF REINFORCEMENT

Steel reinforcing bars are produced in standard lengths controlled by transportability and weight considerations. In general, 60-ft lengths are normally produced. But it is impractical in beams and slabs spanning over several supports to interweave bars of such lengths on site over several spans. Consequently, bars are cut to shorter lengths and lapped at the least critical bending moment locations for bar sizes No. 11 or

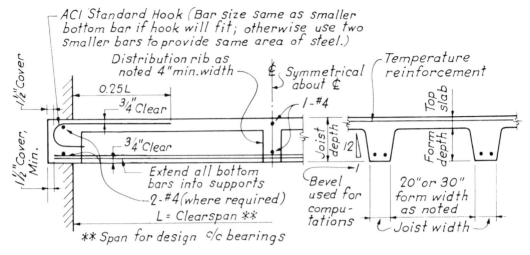

SINGLE SPAN JOIST CONSTRUCTION

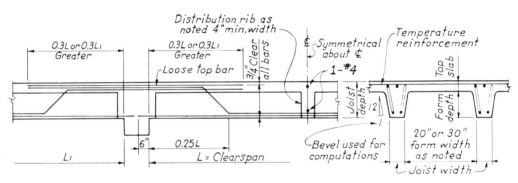

INTERIOR SPAN JOIST CONSTRUCTION

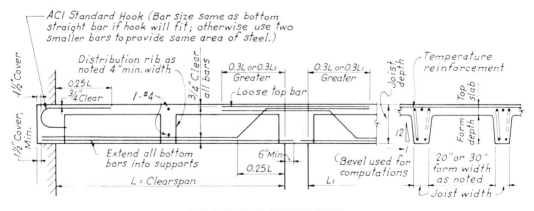

END SPAN JOIST CONSTRUCTION

Figure 4.6 Cutoff point for one-way joist construction. (From Ref. 4.2.)

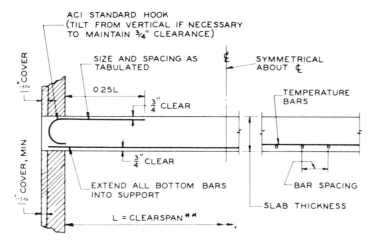

SINGLE SPAN, SIMPLY SUPPORTED

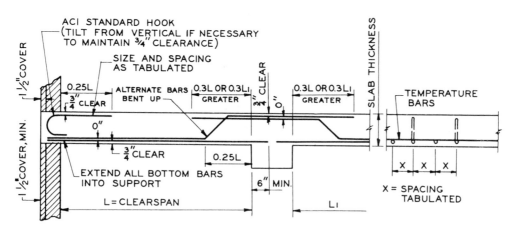

END SPAN, SIMPLY SUPPORTED

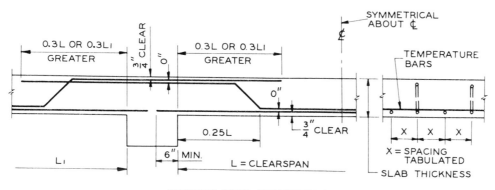

INTERIOR SPAN, CONTINUOUS

Figure 4.7 Cutoff points for one-way slabs. (From Ref. 4.2.)

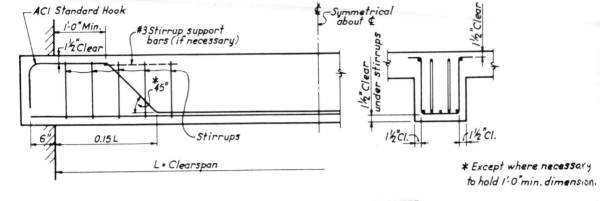

SINGLE SPAN BEAM, SIMPLY SUPPORTED

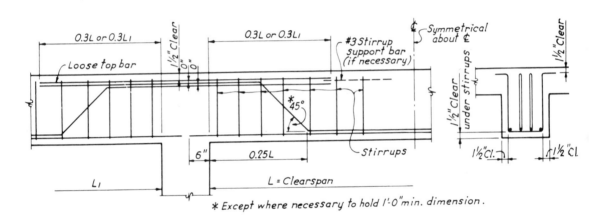

INTERIOR SPAN OF CONTINUOUS BEAM

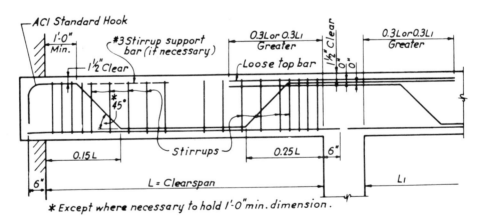

END SPAN OF SIMPLY SUPPORTED BEAM

Figure 4.8 Reinforcing details for continuous beams with diagonal tension steel. (From Ref. 4.2.)

smaller. A general rule of thumb for maximum bar length is about 40 ft for shipping purposes. The most effective means of continuity in reinforcement is to weld the cut pieces without reducing the mechanical or strength properties of the welded bar at the weld. Cost considerations require alternatives, however. Basically there are three types of splicing:

1. *Lap splicing:* It depends on full bond development of the two lapping bars at the lap for bars of size not larger than No. 11.
2. *Welding by fusion of the two bars at the connection:* This process can be economically justifiable for bar sizes larger than No. 11 bars.
3. *Mechanical connecting:* This can be achieved by mechanical sleeves threaded on the ends of the bars to be interconnected. Such connectors should have a yield strength at least 1.25 times the yield strength of the bars they interconnect. They are also more commonly used for large-diameter bars.

4.3.1 Splices of Deformed Bars and Deformed Wires in Tension

Three classes of lap splices are specified by the ACI code. The minimum length l_s but not less than 12 in. is

Class A: $1.0l_{db}$
Class B: $1.3l_{db}$
Class C: $1.7l_{db}$

Table 4.1 gives the maximum percentage of tensile steel area A_s that can be spliced. Splicing should be avoided at maximum tensile stress if at all possible; splicing may be made by simple lapping of bars either in contact or separated by concrete. Every effort should be made to stagger the splices rather than having all the bars spliced within the required lap length.

TABLE 4.1   MAXIMUM TENSION STEEL AREA TO BE SPLICED

$\dfrac{\text{Provided } A_s}{\text{Required } A_s}$	Maximum percentage of A_s spliced within the required lap length		
	60	75	100
≥ 2.0	Class A	Class A	Class B
< 2.0	Class B	Class C	Class C

4.3.2 Splices of Deformed Bars in Compression

The lap length l_s should be equal to at least the development length in compression as given in Section 4.1.3 and Eqs. 4.5a and 4.5b and the modifiers. l_s should also satisfy the following, but not less than 12 in.:

$$f_y \le 60{,}000 \text{ psi} \qquad l_s \ge 0.0005 f_y d_b \qquad (4.9a)$$

$$f_y > 60{,}000 \text{ psi} \qquad l_s \ge (0.0009 f_y - 24) d_b \qquad (4.9b)$$

If the compressive strength f_c' of the concrete is less than 3000 psi, such as might occur in foundations, the splice length l_s must be increased by one-third.

Modifying multipliers with values less than 1.0 are used in heavily reinforced tied compression members (0.83) and in spirally reinforced columns (0.75), but the lap length should not be less than 12 in.

4.3.3 Splices of Deformed Welded Wire Fabric

The minimum lap length l_s measured between the ends of the two lapped fabric sheets of welded deformed wire must be $\ge 1.7 l_d$ or 8 in. (204 mm), whichever is greater. Additionally, the overlap measured between the outermost cross wires of each fabric sheet should not be less than 2 in. (50.8 mm).

SELECTED REFERENCES

4.1   Ferguson, P. M., and J. E. Breen, "Lapped Splices for High Strength Reinforcing Bars," *Journal of the American Concrete Institute*, Proc. Vol. 62, No. 9, September 1965, pp. 1063–1078.

4.2   ACI Committee 315, *ACI Detailing Manual—1980*, Special Publication SP-66, American Concrete Institute, Detroit, 1980, 206 pp.

4.3   ACI Committee 408, "Suggested Development, Splice, and Hook Provisions for Deformed Bars in Tension," ACI 408-1 R79, American Concrete Institute, Detroit, 1979, 3 pp.

4.4   Wire Reinforcement Institute, *Reinforcement Anchorages and Splices*, 3rd ed., WRI Publication, McLean, VA, 1979, 32 pp.

PROBLEMS FOR SOLUTION

4.1   Design the compression lap splice for a column section 16 in. by 16 in. (406 mm by 406 mm) reinforced with eight No. 9 bars (eight bars diameter 28.7 mm) equally spaced around all faces.

(a) $f_c' = 5000$ psi (34.47 MPa)

   $f_y = 60{,}000$ psi (413.7 MPa)

(b) $f_c' = 7000$ psi (48.26 MPa)

   $f_y = 80{,}000$ psi (551.6 MPa)

4.2 Calculate the basic development lengths for the following deformed bars embedded in normalweight concrete.

(a) No. 5, No. 8. Given:

$$f_c' = 5000 \text{ psi } (34.47 \text{ MPa})$$

$$f_y = 60,000 \text{ psi } (413.7 \text{ MPa})$$

(b) No. 14, No. 18. Given:

$$f_c' = 4000 \text{ psi}$$

$$f_y = 60,000 \text{ psi}$$

$$f_y = 80,000 \text{ psi}$$

4.3 Calculate the total embedment length for the bars in Problem 4.1 if they are used as compression reinforcement and the concrete is sand-lightweight.

5

Shear and Torsion

5.1 INTRODUCTION

This chapter presents procedures for the analysis and design of reinforced concrete sections to resist the shear forces resulting from externally applied loads. Because the strength of concrete in tension is considerably lower than its strength in compression, design for shear becomes of major importance in concrete structures.

The behavior of reinforced concrete beams at failure in shear is distinctly different from their behavior in flexure. They fail abruptly without sufficient advanced warning and the diagonal cracks that develop are considerably wider than the flexural cracks. The accompanying photographs show typical beam shear failure in diagonal tension as discussed in subsequent sections. Because of the brittle nature of such failures, the designer should design sections that are adequately strong to resist the external factored shear loads without reaching their shear strength capacity.

5.2 BEHAVIOR OF HOMOGENEOUS BEAMS

Consider the two infinitesimal elements A_1 and A_2 of a rectangular beam in Fig. 5.1a made of homogeneous, isotropic, and linearly elastic material. Figure 5.1b shows the bending stress and shear stress distributions across the depth of the section. The tensile normal stress f_t and the shear stress v are the values in element A_1 across plane a_1–a_1 at a distance y from the neutral axis. Using the principles of classical mechanics, the

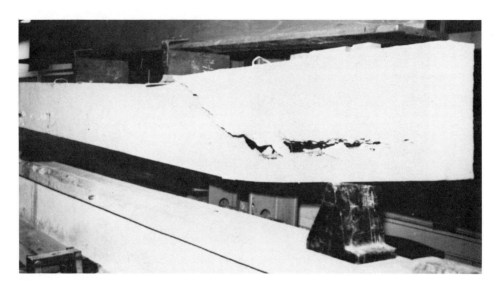

Photo 37   Typical diagonal tension failure at rupture load level. (Tests by Nawy, et al.)

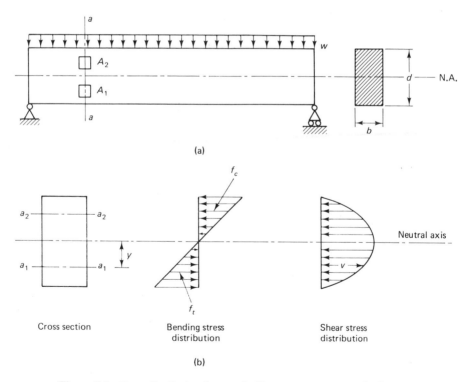

Figure 5.1   Stress distribution for a typical homogeneous rectangular beam.

Photo 38   Simply supported beam prior to developing diagonal tension crack (load stage 11). (Test by Nawy, et al.)

Photo 39   Principal diagonal tension crack at failure of beam in photo 38 (load stage 12).

normal stress f and the shear stress v for element A_1 can be written

$$f = \frac{My}{I} \tag{5.1}$$

and
$$v = \frac{VA\bar{y}}{Ib} \tag{5.2}$$

where M and V = bending moment and shear force at section a_1–a_1

$\quad\quad A$ = cross-sectional area of the section at the plane passing through the centroid of element A_1

$\quad\quad y$ = distance from the element to the neutral axis

$\quad\quad \bar{y}$ = distance from the centroid of A to the neutral axis

$\quad\quad I$ = moment of inertia of the cross section

$\quad\quad b$ = width of the beam

Figure 5.2 shows the internal stresses acting on the infinitesimal elements A_1 and

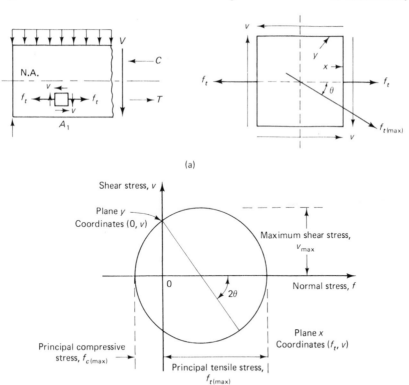

(a)

(b)

Figure 5.2   Stress state in elements A_1 and A_2: (a) stress state in element A_1; (b) Mohr's circle representation, element A_1; (c) stress state in element A_2; (d) Mohr's circle representation, element A_2.

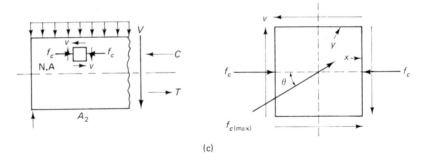

(c)

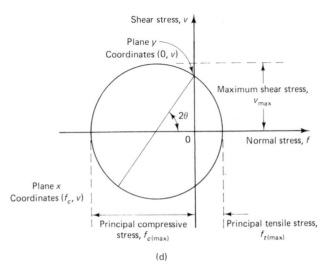

(d)

Figure 5.2 (*cont.*)

A_2. Using Mohr's circle in Fig. 5.2b, the principal stresses for element A_1 in the tensile zone below the neutral axis become

$$f_{t(max)} = \frac{f_t}{2} + \sqrt{\left(\frac{f_t}{2}\right)^2 + v^2} \qquad \text{principal tension} \qquad (5.3a)$$

$$f_{c(max)} = \frac{f_t}{2} - \sqrt{\left(\frac{f_t}{2}\right)^2 + v^2} \qquad \text{principal compression} \qquad (5.3b)$$

and $\qquad \tan 2\theta_{max} = \dfrac{v}{f_t/2} \qquad\qquad\qquad\qquad\qquad\qquad (5.3c)$

5.3 *BEHAVIOR OF REINFORCED CONCRETE BEAMS AS NONHOMOGENEOUS SECTIONS*

The behavior of reinforced concrete beams differs in that the tensile strength of concrete is about one-tenth of its strength in compression. The compression stress f_c in element A_2 of Fig. 5.2b above the neutral axis prevents cracking, for the maximum principal stress in the element is in compression. For element A_1 below the neutral axis, the maximum principal stress is in tension; thus cracking ensues. As one moves toward the support, the bending moment—and hence f_t decreases, accompanied by a corresponding increase in the shear stress. The principal stress $f_{t(max)}$ in tension acts at an approximately 45° plane to the normal at sections close to the support, as seen in Fig. 5.3. Because of the low tensile strength of concrete, diagonal cracking develops along planes perpendicular to the planes of principal tensile stress—hence the term *diagonal tension cracks*. To prevent such cracks from opening, special "diagonal tension" reinforcement must be provided.

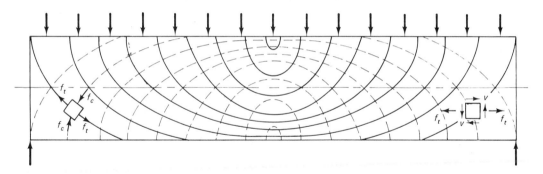

Figure 5.3 Trajectories of principal stresses in a homogeneous isotropic beam. Solid lines: tensile trajectories; dashed lines: compressive trajectories.

If f_t close to the support in Fig. 5.3 is assumed equal to zero, the element becomes nearly in a state of pure shear and the principal tensile stress, using Eq. 5.3b, would be equal to the shear stress v on a 45° plane. It is this diagonal tension stress that causes the inclined cracks.

Definitive understanding of the correct shear mechanism in reinforced concrete is still incomplete. The approach of ACI-ASCE Joint Committee 426, however, gives a systematic empirical correlation of the basic concepts developed from extensive test results.

5.4 DIAGONAL TENSION ANALYSIS OF SLENDER
AND INTERMEDIATE BEAMS

The occurrence of the first inclined crack determines the shear strength of a beam without web reinforcement. Because crack development is a function of the tensile strength of the concrete in the beam web, a knowledge of the principal stress in the critical sections is necessary, as discussed in Sections 5.2 and 5.3. The controlling principal stress in concrete is the result of the shearing stress v_u due to the external factored shear V_u and the horizontal flexural stress f_t due to the external factored bending moment M_u. The ACI code provides an empirical model based on results of extensive tests to failure of a large number of beams without web reinforcement. The model is a regression solution to the basic equation for two-dimensional principal stress at a point.

$$f_{t(max)} = f'_t = \frac{f_t}{2} + \sqrt{\left(\frac{f_t}{2}\right)^2 + v^2} \tag{5.3a}$$

where $f_{t(max)}$ is the principal stress in tension and can be assumed to be equal to a constant multiplied by the tensile splitting strength f'_t of plain concrete. Since f'_t has been proven to be a function of $\sqrt{f'_c}$, Eq. 5.3a becomes

$$\sqrt{f'_c} = K_1 \left[\frac{f_t}{2} + \sqrt{\left(\frac{f_t}{2}\right)^2 + v^2} \right] \tag{5.4}$$

where K_1 is a constant.

The flexural stress f_t in the concrete is a function of the steel stress in the longitudinal reinforcement or the moment of resistance of the section, or

$$f_t \propto \frac{E_c}{E_s} f_s \propto \frac{E_c M_n}{E_s A_s d}$$

But the reinforcement ratio $\rho_w = A_s/bd$ at the tension side and E_c/E_s has a constant value. So

$$f_t = F_1 \frac{M_n}{\rho_w bd^2} \tag{5.5}$$

where F_1 is a constant to be determined by test and M_n is the nominal moment strength of a given section. The shear stress v at the specific cross section bd due to the vertical external factored shear force V_u is

$$v = F_2 \frac{V_n}{bd} \tag{5.6}$$

where V_n is the nominal shear resistance at the section under consideration and F_2 is the other constant, to be determined from the beam tests. Coefficients F_1 and F_2 both depend on several variables, including the geometry of the beam, type of loading, amount and arrangement of reinforcement, and the interaction between the steel reinforcement and the concrete.

Substituting f_t of Eq. 5.5 and v of Eq. 5.6, rearranging terms, and evaluating the constants K_1, F_1, and F_2 of the experimental model yield the following regression expression:

$$\frac{V_n}{bd\sqrt{f'_c}} = 1.9 + 2500\rho_w \frac{V_n d}{M_n \sqrt{f'_c}} \leq 3.5 \tag{5.7}$$

A plot of Eq. 5.7 is shown in Fig. 5.4. It is to be noted that $M_n/V_n d = a/d$; consequently, Eq. 5.7 accounts indirectly for the shear span/depth ratio and thus for the slenderness of the member. If the nominal shear resistance of the *plain* concrete web is termed V_c, then V_n in the left-hand side of Eq. 5.7 must be expressed as V_c. Transforming Eq. 5.7 into a force format for evaluation of the nominal shear resistance of the web of a beam of normal concrete and having no diagonal tension steel gives

$$V_c = 1.9b_w d\sqrt{f'_c} + 2500\rho_w \frac{V_n d}{M_n} b_w d \leq 3.5 b_w d\sqrt{f'_c} \tag{5.8}$$

It is to be emphasized that the ratio $V_u d/M_u$ or $V_n d/M_n$ cannot exceed 1.0, where $V_n = V_u/\phi$ and $M_n = M_u/\phi$ as the values of shear and moment at the section for which V_c is being evaluated.

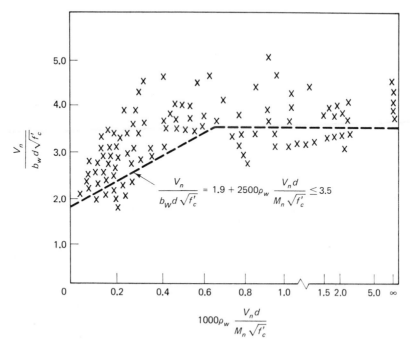

Figure 5.4 Shear resistance of reinforced concrete beam webs.

The first critical values of V_n and M_n are taken at a distance d from the face of the support because the stabilized (principal) diagonal tension cracks develop in that zone. As one moves toward the midspan of the beam, the values of M_n and V_n will change. The appropriate moments M_n and shears V_n must be calculated for the particular section that is being analyzed for web steel reinforcement.

For simplicity of calculations, a more conservative ACI expression can be applied, particularly if the same beam section is not repetitively used in the structure:

$$V_c = \lambda \times 2.0\sqrt{f'_c}\,b_w d \qquad (5.9)$$

where λ is a factor dependent on the type of concrete, with values of 1.0 for normal-weight concrete, 0.85 for sand-lightweight concrete, and 0.75 for all lightweight concrete. When axial compression also exists, V_c in Eq. 5.9 becomes

$$V_c = 2\lambda\left(1 + \frac{N_u}{2000A_g}\right)\sqrt{f'_c}\,b_w d \qquad (5.10a)$$

When significant axial tension exists, then

$$V_c = 2\lambda\left(1 + \frac{N_u}{500A_g}\right)\sqrt{f'_c}\,b_w d \qquad (5.10b)$$

N_u/A_g is expressed in psi, where N_u is the axial load on the member and A_g is the gross area of the section; N_u is negative in tension.

5.5 WEB STEEL PLANAR TRUSS ANALOGY

As noted, web reinforcement must be provided to prevent failure due to diagonal tension. Theoretically if the necessary steel bars in the form of the tensile stress trajectories shown in Fig. 5.3 are placed in the beam, no shear failure can occur. Practical considerations eliminate such a solution, however, and other forms of reinforcing are improvised to neutralize the principal tensile stresses at the critical shear failure planes. The mode of failure in shear reduces the beam to a simulated arched section in compression at the top and tied at the bottom by the longitudinal beam tension bars, as seen in Fig. 5.5a. If one isolates the main concrete compression element shown in Fig. 5.5b, it can be considered as the compression member of a triangular truss, as shown in Fig. 5.5c with the polygon of forces C_c, T_b, and T_s representing the forces acting on the truss members—hence the expression *truss analogy*. Force C_c is the compression in the simulated concrete strut, force T_b is the tensile force increment of the main longitudinal tension bar, and T_s is the force in the bent bar. Figure 5.6a shows the analogy truss for the case of using vertical stirrups instead of inclined bars, with the forces polygon having a vertical tensile force T_s instead of the inclined one in Fig. 5.5c.

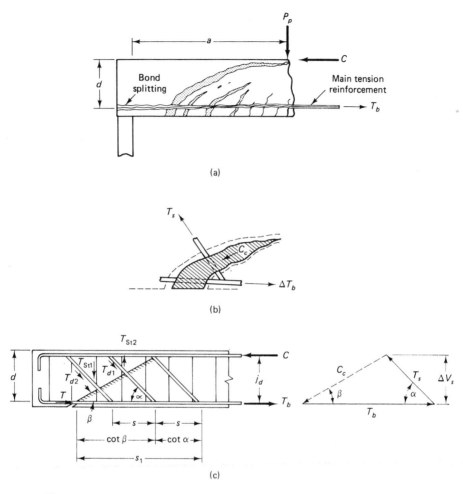

Figure 5.5 Diagonal tension failure mechanism: (a) failure pattern; (b) concrete simulated strut; (c) planar truss analogy.

As can be seen from the preceding discussion, the shear reinforcement basically performs four main functions:

1. Carries a portion of the external factored shear force V_u.
2. Restricts the growth of the diagonal cracks.
3. Holds the longitudinal main reinforcing bars in place so that they can provide the dowel capacity needed to carry the flexural load.
4. Provides some confinement to the concrete in the compression zone if the stirrups are in the form of closed ties.

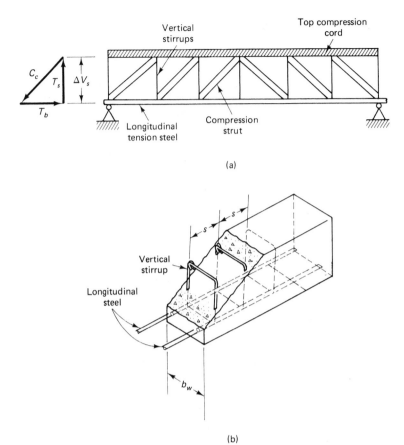

(a)

(b)

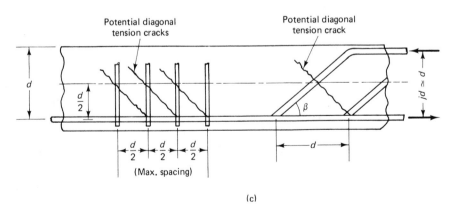

(c)

Figure 5.6 Web steel arrangement: (a) truss analogy for vertical stirrups; (b) three-dimensional view of vertical stirrups; (c) spacing of web steel.

5.5.1 Web Steel Resistance

If V_c, the nominal shear resistance of the plain web concrete, is less than the nominal total vertical shearing force $V_u/\phi = V_n$, then web reinforcement must be provided to carry the difference in the two values; thus

$$V_s = V_n - V_c \qquad (5.11)$$

The nominal resisting shear V_c can be calculated from Eq. 5.8 or 5.9 and V_s can be determined from equilibrium analysis of the bar forces in the analogous triangular truss cell. From Fig. 5.5 one gets

$$V_s = T_s \sin \alpha = C_c \sin \beta \qquad (5.12a)$$

where T_s is the force resultant of all web stirrups across the diagonal crack plane and n is the number of spacings s. If $s_1 = ns$ in the bottom tension chord of the analogous truss cell, then

$$s_1 = jd(\cot \alpha + \cot \beta) \qquad (5.12b)$$

Assuming that moment arm $jd \simeq d$, the stirrup force per unit length from Eqs. 5.12a and 5.12b, where $s_1 = ns$, becomes

$$\frac{T_s}{s_1} = \frac{T_s}{ns} = \frac{V_s}{\sin \alpha} \frac{1}{d(\cot \beta + \cot \alpha)} \qquad (5.12c)$$

If there are n inclined stirrups within the s_1 length of the analogous truss chord, and if A_v is the area of one inclined stirrup, then

$$T_s = nA_v f_y \qquad (5.13a)$$

So

$$nA_v = \frac{V_s ns}{d \sin \alpha(\cot \beta + \cot \alpha)f_y} \qquad (5.13b)$$

But it can be assumed that, in the case of diagonal tension failure, the compression diagonal makes an angle $\beta = 45°$ with the horizontal; Eq. 5.13b becomes

$$V_s = \frac{A_v f_y d}{s} [\sin \alpha(1 + \cot \alpha)]$$

to get

$$V_s = \frac{A_v f_y d}{s}(\sin \alpha + \cos \alpha) \qquad (5.14a)$$

or

$$s = \frac{A_v f_y d}{V_n - V_c}(\sin \alpha + \cos \alpha) \qquad (5.14b)$$

If the inclined web steel consists of a single bar or a single group of bars all bent at the same distance from the face of the support,

$$V_s = A_v f_y \sin \alpha \le 3.0\sqrt{f_c'}\, b_w d$$

If vertical stirrups are used, angle α becomes 90°, giving

$$V_s = \frac{A_v f_y d}{s} \tag{5.15a}$$

or

$$s = \frac{A_v f_y d}{(V_u/\phi) - V_c} = \frac{A_v \phi f_y d}{V_u - \phi V_c} \tag{5.15b}$$

5.5.2 Limitations on Size and Spacing of Stirrups

Equations 5.14 and 5.15 give inverse relationships between the spacing of the stirrups and the shear force or shear stress they resist, with the spacing s decreasing with the increase in $(V_n - V_c)$. In order for every *potential* diagonal crack to be resisted by a vertical stirrup as seen in Fig. 5.5c, maximum spacing limitations are to be applied as follows for vertical stirrups:

1. $V_n - V_c > 4\sqrt{f_c'}\, b_w d$: $s_{max} = d/4 \le 24$ in.
2. $V_n - V_c \le 4\sqrt{f_c'}\, b_w d$: $s_{max} = d/2 \le 24$ in.
3. $V_n - V_c > 8\sqrt{f_c'}\, b_w d$: enlarge section

The minimum web steel area A_v must be provided if the factored shear force V_u exceeds one-half the shear strength ϕV_c of the plain concrete web. This precaution is necessary to prevent brittle failure, thus enabling both the stirrups and the beam compression zone to continue carrying the increasing shear after the formulation of the first inclined crack.

$$\text{Minimum } A_v = \frac{50 b_w s}{f_y} \tag{5.16}$$

where A_v is the area of all the vertical stirrup legs in the cross section.

5.6 WEB REINFORCEMENT DESIGN PROCEDURE FOR SHEAR

Here is a summary of the recommended sequence of design steps.

1. Determine the critical section and calculate the factored shear force V_u. When the reaction, in the direction of applied shear, introduces compression into the end regions of a member, the critical section can be assumed at a distance of d from the support, provided that no concentrated load acts between the support face and distance d thereafter.

2. Check whether

$$V_u \leq \phi(V_c + 8\sqrt{f'_c}\, b_w d)$$

where b_w is the web width. If this condition is not satisfied, the cross section has to be enlarged.

3. Use minimum shear reinforcement A_v if V_u is larger than one-half ϕV_c, with the following exceptions:

(a) Concrete joist construction
(b) Slabs and footings
(c) Small shallow beams of depth not exceeding 10 in. (254 mm) or $2\frac{1}{2}$ times the flange thickness:

$$\text{Minimum } A_v = \frac{50 b_w s}{f_y}$$

Good construction practice dictates that some stirrups always be used to facilitate proper handling of the reinforcement cage.

4. If $V_u > \phi V_c$, shear reinforcement must be provided such that $V_u \leq \phi(V_c + V_s)$, where

$$V_s = \begin{cases} \dfrac{A_v f_y d}{s} & \text{for vertical strirrups} \\[2ex] \dfrac{A_v f_y d}{s}(\sin \alpha + \cos \alpha) & \text{for inclined stirrups} \end{cases}$$

5. Maximum spacing s must be $s = d/2 \leq 24$ in., except that in cases where $V_s > 4\sqrt{f'_c}\, b_w d$, the spacing then becomes $s \leq d/4 \leq 24$ in.

Figure 5.7 presents a flowchart for the performance of the sequence of calculations necessary for the design of vertical stirrups. Simple corresponding modifications of this chart can be made so that the chart can be used in the design of inclined web steel.

5.7 EXAMPLES ON THE DESIGN OF WEB STEEL FOR SHEAR

5.7.1 Example 5.1: Design of Web Stirrups

A rectangular isolated beam has an effective span of 25 ft (7.62 m) and carries a working live load of 8000 lb per linear foot (10.85 kN/m) and no external dead load except its self-weight. Design the necessary shear reinforcement. Use the simplified term of Eq. 5.9 for calculating the capacity V_c of the plain concrete web. Given:

$f'_c = 4000$ psi (27.6 MPa), normalweight concrete
$f_y = 60,000$ psi (414 MPa)

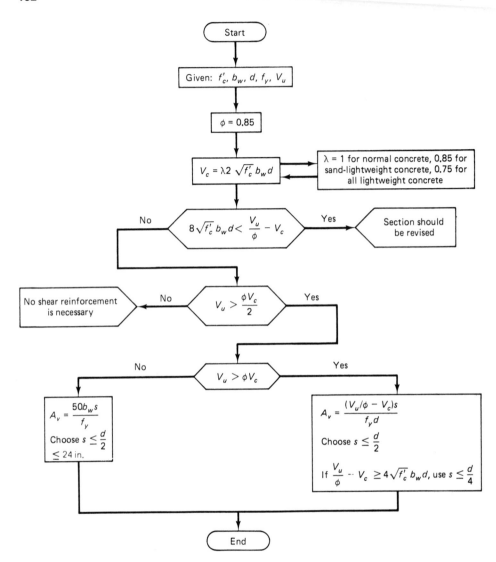

Figure 5.7 Flowchart for web reinforcement design procedure.

$b_w = 14$ in. (356 mm)

$d = 28$ in. (712 mm)

$h = 30$ in. (762 mm)

Longitudinal tension steel: six No. 9 bars (diameter 28.6 mm)

No axial force acts on the beam.

Solution

Factored shear force (Step 1)

$$\text{Beam self-weight} = \frac{14 \times 30}{144} \times 150 = 437.5 \text{ lb/ft}$$

$$\text{Total factored load} = 1.7 \times 8000 + 1.4 \times 437.5 = 14,212.5 \text{ lb/ft}$$

The factored shear force at the face of the support is

$$V_u = \frac{25}{2} \times 14,212.5 = 177,656 \text{ lb}$$

The first critical section is at a distance $d = 28$ in. from the face of the support of this beam (half-span $= 150$ in.)

$$V_u \text{ at } d = \frac{150 - 28}{150} \times 177,656 = 144,494 \text{ lb}$$

Shear capacity (Step 2)

The shear capacity of the plain concrete in the web from the simplified equation for normalweight concrete ($\lambda = 1.0$) is

$$V_c = 2.0\lambda\sqrt{f'_c}\, b_w d = 2 \times 1.0\sqrt{4000} \times 14 \times 28 = 49,585 \text{ lb}$$

Check for adequacy of section for shear:

$$(8 + 2.0)\sqrt{f'_c}\, b_w d = 10\sqrt{f'_c}\, b_w d = 247,923 \text{ lb}$$

$$V_n = \frac{V_u}{\phi} = \frac{144,494}{0.85} = 169,993 \text{ lb} \qquad \text{cross section O.K.}$$

$$V_n > V_c \qquad \text{Hence stirrups are necessary.}$$

Shear reinforcement (Steps 3 to 5)

Try No. 4 two-legged stirrups (area per leg $= 0.20$ in.^{2}).

$$A_v = 2 \times 0.2 = 0.40 \text{ in.}^2$$

From Eq. 5.15b

$$s = \frac{A_v f_y d}{(V_u/\phi) - V_c} = \frac{0.4 \times 60,000 \times 28}{169,993 - 49,584}$$

$$= 5.58 \text{ in. } (141.7 \text{ mm})$$

Because $V_n - V_c > 4\sqrt{f'_c}\, b_w d$, the maximum allowable spacing $s = d/4 = 28/4 = 7$ in. At the critical section, $d = 28$ in. from the face of the support, the maximum allowable spacing would be 5.58 in. in this case.

The shear force for distributed load decreases linearly from the support to midspan of the beam. So the web reinforcement can be reduced accordingly after determining the zone where minimum reinforcement is necessary and the zone where no web reinforcement is needed. The same size and spacing of stirrups needed at the critical section d from face of support should be continued to the support. Figure 5.8 illustrates the various values being calculated.

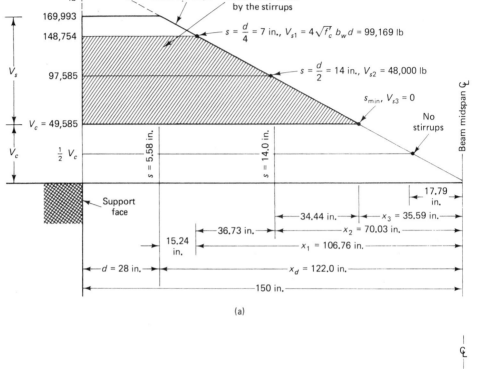

(a)

(b)

Figure 5.8 Stirrups arrangements for Ex. 5.1: (a) shear envelope and stirrup design segments; (b) vertical stirrups spacing.

Critical Phase x_d (consider the midspan as the origin): $V_n = 169,993$ lb and, as noted, $s = 5.58$ in. x_d from the midspan point $= 150 - 28 = 122$ in.

Plane x_1 at $s = d/4$ *maximum spacing:*

$$V_{s1} = 4\sqrt{f'_c}\, b_w d = 4\sqrt{4000} \times 14 \times 28 = 99,169 \text{ lb}$$

$$V_{n1} = 99,169 + 49,585 = 148,754 \text{ lb}$$

$$x_1 \text{ from midspan point} = (150 - 28) \times \frac{148,754}{169,993} = 106.76 \text{ in.}$$

Plane x_2 at $s = d/2$ maximum spacing:

$$s = \frac{A_v f_y d}{V_n - V_c} \quad \text{or} \quad \frac{28}{2} = \frac{0.4 \times 60,000 \times 28}{V_s}$$

or
$$V_{s2} = 48,000 \text{ lb}$$

$$V_{n2} = 48,000 + 49,585 = 97,585 \text{ lb}$$

$$x_2 \text{ from midspan point} = 122 \times \frac{97,585}{169,993} = 70.03 \text{ in.}$$

From Fig. 5.8a the distance 36.73 in. is the transition zone from $s = 7$ in. to $s = 14$ in.; thus a stirrup spacing of 8 in. center to center is shown in Fig. 5.8b.

Plane x_3 at shear force V_c:

$$V_c = 2\sqrt{f_c'}\, b_w d = 49,585 \text{ lb}$$

$$x_3 \text{ from midspan point} = 122 \times \frac{49,585}{169,993} = 35.59 \text{ in.}$$

Extend stirrups to a distance $d = 28$ in. beyond the point where they are theoretically not required.

Minimum web steel: Test when $V_u > \frac{1}{2}\phi V_c$ or $V_n > \frac{1}{2}V_c$

$$V_u = 144,494$$

$$\frac{1}{2}V_c = \frac{1}{2} \times 49,585 = 24,793 \text{ lb}$$

$$A_v = \frac{50 b_w s}{f_y} = \frac{50 \times 14 \times 14}{60,000} = 0.16 \text{ in.}^2$$

$$< \text{actual } A_v = 0.40 \text{ in.}^2 \quad \text{O.K.}$$

or
$$\text{Maximum required } s = \frac{A_v f_y}{50 b_w} = \frac{0.40 \times 60,000}{50 \times 14} = 34.3 \text{ in.}$$

$$\text{versus maximum used } s = \frac{d}{2} = 14 \text{ in.} \quad \text{O.K.}$$

$$x_y = 122.0 \times \frac{24,793}{169,993} = 17.79 \text{ in. from midspan}$$

Proportion the spacing of the vertical stirrups accordingly.

The shaded area in Fig. 5.8a is the shear force area for which stirrups must be provided. The spacing of the stirrups in Fig. 5.8b is based on the practical consideration of the desirability to use whole spacing dimensions and to vary the spacing as little as possible.

5.8 TORSION THEORY

Torsion occurs in monolithic concrete construction primarily where the load acts at a distance from the longitudinal axis of the structural member. An end beam in a floor panel, a spandrel beam receiving load from one side, a canopy or a bus-stand roof projecting from a monolithic beam on columns, peripheral beams surrounding a floor opening, or a helical staircase are all examples of structural elements subjected to twisting moments. These moments occasionally cause excessive shearing stresses. As a result, severe cracking can develop well beyond the allowable serviceability limits unless special torsional reinforcement is provided. Photos 40 and 41 illustrate the extent of cracking at failure of a beam in torsion. They show the curvilinear plane of twist caused by the imposed torsional moments. In actual spandrel beams of a structural system, the extent of damage due to torsion is usually not as severe, as seen in photos 42 and 43. This situation is due to the redistribution of stresses in the structure. Loss of integrity due to torsional distress, however, should always be avoided by proper design of the necessary torsional reinforcement.

An introduction to the subject of torsional stress distribution must start with the basic elastic behavior of simple sections, such as circular or rectangular sections. Most concrete beams subjected to twist are components of rectangles. They are usually flanged sections, such as T beams and L beams. Although circular sections are rarely a consideration in normal concrete construction, a brief discussion of torsion in circular sections serves as a good introduction to the torsional behavior of other types of sections.

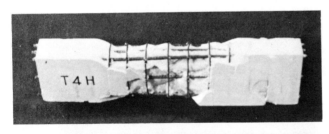

Photo 40 Reinforced plaster beam at failure in pure torsion. (Rutgers tests: Law, Nawy, et al.)

(a)

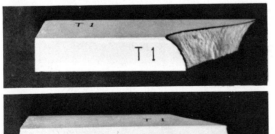

(b)

Photo 41 Plain mortar beam in pure torsion: (a) top view; (b) bottom view. (Rutgers tests: Law, Nawy, et al.)

Photo 42   Reinforced concrete beams in torsion—testing setup. (Courtesy of Thomas T. C. Hsu.)

Photo 43   Closeup of torsional cracking of beams in photo 42. (Courtesy of Thomas T. C. Hsu.)

Shear stress is equal to shear strain times the shear modulus at the elastic level in circular sections. As in the case of flexure, the stress is proportional to its distance from the neutral axis (i.e., the center of the circular section) and is maximum at the extreme fibers. If r is the radius of the element, $J = \pi r^4/2$, its polar moment of inertia, and v_{te} the elastic shearing stress due to an elastic twisting moment T_e,

$$v_{te} = \frac{T_e r}{J} \tag{a}$$

When deformation takes place in the circular shaft, the axis of the circular cylinder is assumed to remain straight. All radii in a cross section also remain straight (i.e., without warping) and rotate through the same angle about the axis. As the circular element starts to behave plastically, the stress in the plastic outer ring becomes constant whereas the stress in the inner core remains elastic, as shown in Fig. 5.9. As the whole cross section becomes plastic, $b = 0$ and the shear stress

$$v_{tf} = \frac{3}{4} \frac{T_p r}{J} \tag{b}$$

where v_{tf} is the nonlinear shear stress due to an ultimate twisting moment T_p, where the subscript f denotes failure.

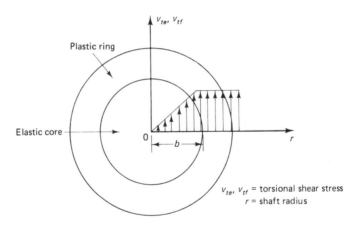

Figure 5.9   Torsional stress distribution through circular section.

The torsional problem is considerably more complicated in rectangular sections. The originally plane cross sections undergo warping due to the applied torsional moment. This moment produces axial as well as circumferential shear stresses with zero values at the corners of the section and the centroid of the rectangle, and maximum values on the periphery at the middle of the sides, as seen in Fig. 5.10. The maximum torsional shearing stress would occur at midpoints A and B of the larger dimension of the cross section. These complications, plus the fact that reinforced concrete sections are neither homogeneous nor isotropic, make it difficult to develop

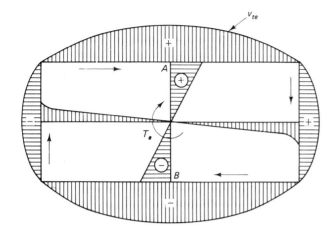

Figure 5.10 Pure torsion stress distribution in a rectangular section.

exact mathematical formulations based on physical models, such as Eqs. (a) and (b) for circular sections.

For over 60 years the torsional analysis of concrete members has been based on either (a) the classical theory of elasticity developed through mathematical formulations coupled with membrane analogy verifications (St.-Venant's) or (b) the theory of plasticity represented by the sand-heap analogy (Nadai's). Both theories were applied essentially to the state of pure torsion. But it was found experimentally that the elastic theory is not entirely satisfactory for the accurate prediction of the state of stress in concrete in pure torsion. The behavior of concrete was found to be better represented by the plastic approach. Consequently, almost all developments in torsion as applied to concrete and reinforced concrete have been in the latter direction.

5.9 TORSION IN REINFORCED CONCRETE ELEMENTS

Torsion rarely occurs in concrete structures without being accompanied by bending and shear. The capacity of the plain concrete to resist torsion when in combination with other loads could, in many cases, be lower than when it resists the same factored external twisting moments alone. Consequently, torsional reinforcement must be provided to resist the excess torque.

Inclusion of longitudinal and transverse reinforcement to resist part of the torsional moments introduces a new element in the set of forces and moments in the section. If

T_n = required total nominal torsional resistance of the section, including the reinforcement

T_c = nominal torsional resistance of the plain concrete

T_s = torsional resistance of the reinforcement

then

$$T_n = T_c + T_s \qquad (5.17a)$$

or

$$T_s = T_n - T_c \qquad (5.17b)$$

5.10 DESIGN OF REINFORCED CONCRETE BEAMS SUBJECTED TO COMBINED TORSION, BENDING, AND SHEAR

5.10.1 Torsional Behavior of Structures

The torsional moment acting on a particular structural component, such as a spandrel beam, can be calculated by using normal structural analysis procedures. Design of the particular component needs to be based on the limit state at failure. Therefore the nonlinear behavior of a structural system after torsional cracking must be identified in one of the following two conditions: (a) no redistribution of torsional stresses to other members after cracking and (b) redistribution of torsional stresses and moments after cracking to effect deformation compatibility between intersecting members.

Stress resultants due to torsion in statically determinate beams can be evaluated from equilibrium conditions alone. Such conditions require a design for the full factored external torsional moment, for no redistribution of torsional stresses is possible. This state is often termed *equilibrium torsion*. An edge beam supporting a cantilever canopy as in Fig. 5.11 is such an example.

The edge beam must be designed to resist the *total* external factored twisting moment due to the cantilever slab; otherwise the structure will collapse. Failure would be caused by the beam not satisfying conditions of equilibrium of forces and moments resulting from the large external torque.

In statically indeterminate systems, stiffness assumptions, compatibility of strains at the joints, and redistribution of stresses may affect the stress resultants,

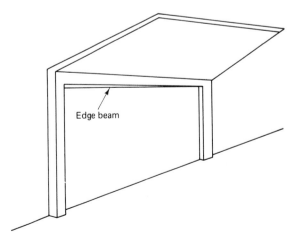

Edge beam

Figure 5.11 No redistribution torsion (equilibrium torsion).

leading to a reduction of the resulting torsional shearing stresses. A reduction is permitted in the value of the factored moment used in the design of the member if part of this moment can be redistributed to the intersecting members. The ACI code allows a maximum factored torsional moment at the critical section d from the face of the supports:

$$T_u = \phi\left(4\sqrt{f_c'}\,\frac{\Sigma\,x^2y}{3}\right) \tag{5.18}$$

$\Sigma x^2 y$ represents the polar moment of inertia of the section where x is the smaller and y is the larger external dimension of the rectangular section. A T-section is broken into three rectangular sections as shown subsequently in Fig. 5.14. An arithmetical summation is taken of the moments of inertia of the three component rectangles to arrive at the total value.

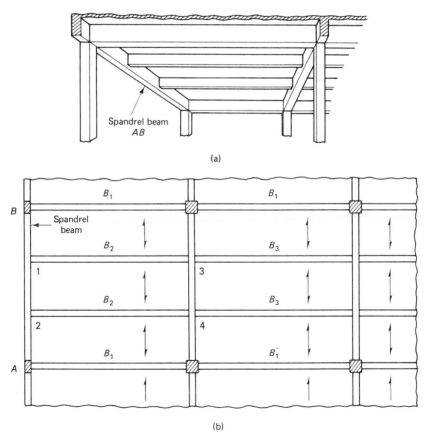

(a)

(b)

Figure 5.12 Torsion redistribution (compatibility): (a) isometric view of one end panel; (b) plan of a typical one-way floor system.

Neglect of the full effect of the total external torsion in this case does not, in effect, lead to failure of the structure but may result in excessive cracking if $\phi(4\sqrt{f_c'}\ \Sigma\ x^2y/3)$ is considerably smaller in value than the actual factored torque. An example of compatibility torsion can be seen in Fig. 5.12.

Beams B_2 apply twisting moments T_u at sections 1 and 2 of spandrel beam AB in Fig. 5.12b. The magnitudes of relative stiffnesses of beam AB and transverse beams B_2 determine the magnitudes of rotation at intersecting joints 1 and 2. Because of continuity and two-way action, the end moments for beams B_2 at their intersections with spandrel beam AB will not be fully transferred as twisting moments to the column supports at A and B. They would be greatly reduced as moment redistribution results in transfer of most of the end bending moments from end 1 to end 3 and end 2 to end 4 of beams B_2. T_u at each spandrel beam supports A and B and at the critical section a distance d from these supports is determined from Eq. 5.18:

$$T_u = \phi\left(4\sqrt{f_c'}\ \frac{\Sigma\ x^2y}{3}\right)$$

If the actual factored torque due to beams B_2 is less than that given by Eq. 5.18, the beam must be designed for the lesser torsional value. Torsional moments are neglected, however, if

$$T_u < \phi\left(0.5\sqrt{f_c'}\ \Sigma\ x^2y\right) \tag{5.19}$$

When the factored torsional moment T_u exceeds $\phi(0.5\sqrt{f_c'}\ \Sigma\ x^2y)$, the ACI code requires that the plain concrete web in the sections be designed for

$$V_c = \frac{2\sqrt{f_c'}\ b_w d}{\sqrt{1 + [2.5C_t(T_u/V_u)]^2}} \tag{5.20a}$$

and

$$T_c = \frac{0.8\sqrt{f_c'}\ \Sigma\ x^2y}{\sqrt{1 + (0.4V_u/C_tT_u)^2}} \tag{5.20b}$$

Equations 5.20a and 5.20b are derived by assuming that the ratio of the torsional moment to the shear force remains constant throughout the loading history. When the contribution of torsion reinforcement is taken into account, the ACI limits the torsional force T_s resisted by the steel to a value not exceeding $4T_c$.

5.10.2 Torsional Web Reinforcement

Meaningful additional torsional strength due to the addition of torsional reinforcement can be achieved only by using *both* stirrups and longitudinal bars. Ideally, equal volumes of steel in both the closed stirrups and the longitudinal bars should be used so that both participate equally in resisting the twisting moments. This principle is the basis of the ACI expressions for proportioning the torsional web steel. If s is the spacing of the stirrups, A_l is the total cross-sectional area of the longitudinal bars, and A_t is the cross section of one stirrup leg, where the dimensions of the stirrup are x_1 in

Photo 44 Terminal building, Dallas International Terminal. (Courtesy of Ammann & Whitney.)

the short direction and y_1 in the long direction, then

$$2A_t(x_1 + y_1) = A_l s \qquad (5.21\text{a})$$

so that

$$2A_t = \frac{A_l s}{x_1 + y_1} \qquad (5.21\text{b})$$

Thus the total torsional web steel, including both the closed stirrups and the longitudinal bars for Eqs. 5.21a and 5.21b, becomes

$$A_{\text{total}} = 2A_t + \frac{A_l s}{x_1 + y_1} \qquad (5.22\text{a})$$

But from Eq. 7.12

$$A_t = \frac{T_s s}{\alpha_1 x_1 y_1 f_y} \qquad (5.22\text{b})$$

where $\alpha_1 = 0.66 + 0.33 y_1/x_1 \le 1.5$ and T_s is the torsional resisting moment of the torsional web steel. If T_c is the nominal torsional resistance of the plain concrete in web,

$$T_s = T_n - T_c \qquad (5.23)$$

From Eq. 5.21b and using the ACI expression for A_t for combined torsion and shear, where

$$2A_t = \frac{200 \times s}{f_y} \frac{T_u}{T_u + V_u/3C_t}$$

the longitudinal torsional reinforcement can be expressed as

$$A_l = \left(\frac{400 \times s}{f_y} \frac{T_u}{T_u + V_u/3C_t} - 2A_t \right) \frac{x_1 \times y_1}{s} \tag{5.24}$$

where $C_t = b_w d / \Sigma\, x^2 y$. The term $2A_t$ in Eq. 5.24 cannot be less than $50\, b_w s / f_y$, for this value is the minimum $2A_t$ for the torsional stirrups to be effective. A thorough discussion and detailed derivation of Eq. 5.24 are presented in Ref. 5.1.

A reduction in stirrups can be compensated by an increase in longitudinal steel as long as the total torsional steel *volume* remains the same. If the spacing s of the stirrups is small such that $2A_t$ is considerably larger than the minimum value $50 b_w s / f_y$, it is not uncommon that A_l from Eq. 5.24 gives a negative value so that the minimum A_l from Eq. 5.21a for equal volumes of stirrups and longitudinal bars is invoked; that is,

$$A_l = 2A_t \frac{x_1 + y_1}{s} \tag{5.25}$$

The total area A_{vt} of the closed stirrups for combined torsion and shear becomes

$$A_{vt} = 2A_t + A_v \geq \frac{50 b_w s}{f_y} \tag{5.26}$$

5.10.3 Design Procedure for Combined Torsion and Shear

The following list is a summary of the recommended sequence of design steps. A flowchart describing the sequence of operations in graphical form is shown in Fig. 5.13.

1. Classify whether the applied torsion is equilibrium or compatibility torsion. Determine the critical section and calculate the factored torsional moment T_u. The critical section is taken at a distance d from the face of the support. If T_u is less than $\phi(0.5\sqrt{f_c'}\,\Sigma\, x^2 y)$, torsional effects can be neglected.
2. Calculate the nominal torsional resistance T_c of the plain concrete web:

$$T_c = \frac{0.8\sqrt{f_c'}\,\Sigma\, x^2 y}{\sqrt{1 + (0.4V_u / C_t T_u)^2}}$$

where $C_t = b_w d / \Sigma\, x^2 y$. Members subjected to significant axial tension may be designed for a T_c value that is multiplied by $(1 + N_u/500Ag)$, where N_u is negative for tension.

Check if T_u exceeds ϕT_c. If it does not, disregard torsional effect. If it does, calculate the value T_s of that portion of the twisting moment to be resisted by the steel reinforcement. For equilibrium torsion,

$$T_s = T_n - T_c$$

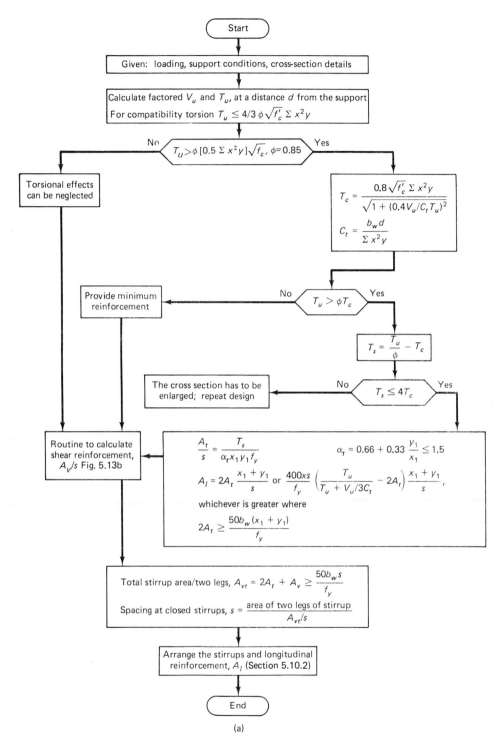

Figure 5.13 Flowchart for the design reinforcement for combined shear and torsion: (a) torsional web steel; (b) shear web steel.

145

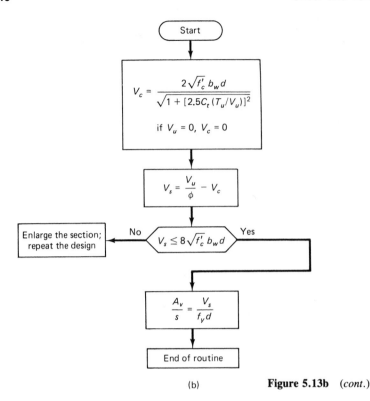

(b)   **Figure 5.13b**   (*cont.*)

For compatibility torsion,

$$T_s = \frac{4}{3}\sqrt{f'_c} \sum x^2 y - T_c \quad \text{or} \quad T_s = T_n - T_c$$

whichever is less. The value of T_n should be at least equivalent to T_u/ϕ. If $T_s > 4T_c$, enlarge the section.

Select the closed stirrups to be used as transverse reinforcement. A minimum No. 3 (9.5 mm diameter) bar size can be used. If $s =$ constant spacing of the stirrups, calculate the area of torsion stirrup per *one* leg per unit spacing:

$$\frac{A_t}{s} = \frac{T_s}{\alpha_t x_1 y_1 f_y}$$

3. Calculate the required shear reinforcement A_v per unit spacing in a transverse section. V_u is the factored external shear force at the critical section, V_c is the nominal shear resistance of the concrete in the web, and V_s is the shearing force to be resisted by the stirrups:

$$\frac{A_v}{s} = \frac{V_s}{f_y d}$$

where $V_s = V_n - V_c$ and

$$V_c = \frac{2\sqrt{f'_c}\, b_w d}{\sqrt{1 + [2.5C_t(T_u/V_u)]^2}}$$

The value of V_n should be at least equal to V_u/ϕ.

4. Obtain the total A_{vt}, the area of closed stirrups for torsion and shear, and design the stirrups such that

$$A_{vt} = 2A_t + A_v \geq \frac{50b_w s}{f_y}$$

5. Calculate the area of longitudinal reinforcement A_l required for torsion where

$$A_l = 2A_t \frac{x_1 + y_1}{s}$$

or $$A_l = \left(\frac{400xs}{f_y}\frac{T_u}{T_u + V_u/3C_t} - 2A_t\right)\frac{x_1 + y_1}{s}$$

whichever is larger. A_l calculated by using the second expression need not exceed

$$A_l = \left(\frac{400xs}{f_y}\frac{T_u}{T_u + V_u/3C_t} - \frac{50b_w s}{f_y}\right)\frac{x_1 + y_1}{s}$$

6. Arrange the reinforcement, using the following guidelines.
   (a) Spacing s of closed stirrups should be less than $(x_1 + y_1)/4$ or 12 in.
   (b) Longitudinal bars should be equally spaced around the perimeter of the closed stirrups. The distance between the bars should be less than 12 in. and at least one longitudinal bar should be placed in each corner.
   (c) Design yield strength of torsion reinforcement should not exceed 60,000 psi.
   (d) Stirrups used for torsion reinforcement should be anchored through a distance d from the extreme compression fibers. Closed ties with hooks at ends achieve this effect.
   (e) Torsional reinforcement should be provided at least a distance $(d + b)$ beyond the point theoretically required in order to cover any potential excessive shearing stresses.

5.10.4 Example 5.2: Design of Web Reinforcement for Combined Torsion and Shear in a T-Beam Section

A T-beam cross section has the geometrical dimensions shown in Fig. 5.14. A factored external shear force acts at the critical section, having a value $V_u = 15{,}000$ lb (67.5 kN). It is subjected to the following torques: (a) equilibrium factored external torsional

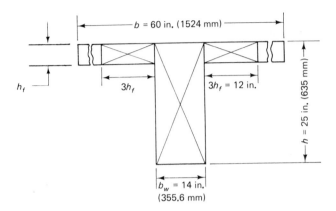

Figure 5.14 Component rectangles of the T beam.

moment T_u = 500,000 in.-lb (57.15 kN-m), (b) compatibility factored T_u = 75,000 in.-lb (8.47 kN-m), and (c) compatibility factored T_u = 300,000 in.-lb. Given:

Bending reinforcement A_s = 3.4 in.^{2} (2193 mm^2)

f_c' = 4000 (27.58 MPa), normalweight concrete

f_y = 60,000 (413.7 MPa)

Design the web reinforcement needed for this section.

Solution

(a) Equilibrium torsion:

Factored torsional moment (Step 1)

Given equilibrium torsional moment = 500,000 in.-lb (57.15 kN-m). The total torsional moment must be provided for in the design. From Fig. 5.14.

$$\Sigma x^2 y = 14^2 \times 25 + 4^2 \times 3 \times 4 + 4^2 \times 3 \times 4 = 5284 \text{ in.}^3$$

$$\phi(0.5\sqrt{f_c'}\ \Sigma\ x^2 y) = 0.85 \times 0.5 \times \sqrt{4000} \times 5284 = 142{,}030 \text{ in.-lb} < T_u$$

Therefore stirrups should be provided.

Torsion closed stirrup design (Step 2)

$$T_n = \frac{T_u}{\phi} = \frac{500{,}000}{0.85} = 588{,}235 \text{ in.-lb (66.47 kN-m)}$$

$$T_c = \frac{0.8\sqrt{f_c'}\ \Sigma\ x^2 y}{\sqrt{1 + (0.4V_u/C_tT_u)^2}}$$

Assume an effective cover of 2.5 in. and d = 25.0 − 2.5 = 22.5 in.

$$C_t = \frac{b_w d}{\Sigma\ x^2 y} = \frac{14 \times 22.5}{5284} = 0.0596$$

$$T_c = \frac{0.8\sqrt{4000} \times 5284}{\sqrt{1 + \left(\dfrac{0.4 \times 15,000}{0.0596 \times 500,000}\right)^2}} = 262,092 \text{ in.-lb (29.61 kN-m)}$$

Also assume that both T_c and V_c are constant for all practical purposes to the midspan of this beam.

$$T_s = T_n - T_c = 588,235 - 262,092 = 326,143 \text{ in.-lb (36.85 kN-m)}$$

Assume $1\frac{1}{2}$ in. clear cover and No. 4 closed stirrups.

$$x_1 = 14 - 2(1.5 + 0.25) = 10.5 \text{ in.}$$

$$y_1 = 25 - 2(1.5 + 0.25) = 21.5 \text{ in.}$$

$$\alpha_t = 0.66 + 0.33 \times \frac{21.5}{10.5} = 1.34 < 1.5$$

Use $\alpha_t = 1.34$.

$$\frac{A_t}{s} = \frac{T_s}{f_y \alpha_t x_1 y_1} = \frac{326,143}{60,000 \times 1.34 \times 10.5 \times 21.5}$$

$$= 0.0180 \text{ in.}^2/\text{in. spacing/one leg}$$

Shear stirrup design (Step 3)

$$V_c = \frac{2\sqrt{f_c'}\, b_w d}{\sqrt{1 + [2.5C_t(T_u/V_u)]^2}} = \frac{2\sqrt{4000} \times 14 \times 22.5}{\sqrt{1 + (2.5 \times 0.0596 \times 500,000/15,000)^2}}$$

$$= 7865 \text{ lb (35.39 kN)}$$

$$V_s = V_n - V_c = \frac{15,000}{0.85} - 7865 = 9782 \text{ lb (44.2 kN)}$$

$$\frac{A_v}{s} = \frac{V_s}{f_y d} = \frac{9782}{60,000 \times 22.5} = 0.0072 \text{ in.}^2/\text{in. spacing/two legs}$$

Combined closed stirrups for torsion and shear (Step 4)

$$\frac{A_{vt}}{s} = \frac{2A_t}{s} + \frac{A_v}{s} = 2 \times 0.0180 + 0.0072 = 0.0432 \text{ in.}^2/\text{in. /two legs}$$

Try No. 3 (9.5 mm diameter) closed stirrups. The area for two legs = 0.22 in.^{2} (142 mm^2).

$$s = \frac{\text{area of stirrup cross section}}{\text{required } A_{vt}/s} = \frac{0.22}{0.0432} = 5.09 \text{ in.}$$

Maximum allowable spacing, s_{max}

$$= \frac{x_1 + y_1}{4} = \frac{10.5 + 21.5}{4} = 8 \text{ in.} > 5.09 \text{ in.} \qquad \text{O.K.}$$

Use No. 3 closed stirrups at 5 in. (127 mm) center to center.

$$\text{Minimum stirrups area required} = A_v + 2A_t = \frac{50b_w s}{f_y} = \frac{50 \times 14 \times 5}{60,000} = 0.0583 \text{ in.}^2$$

$$\text{Area provided} = 0.22 \text{ in.}^2 > 0.0583 \text{ in.}^2 \quad \text{O.K.}$$

Longitudinal torsion steel design (Step 5)

$$A_l = 2A_t \frac{x_1 + y_1}{s} = 2 \times 0.018(10.5 + 21.5) = 1.152 \text{ in.}^2$$

Also,

$$A_l = \left(\frac{400xs}{f_y} \frac{T_u}{T_u + V_u/3C_t} - 2A_t \right) \frac{x_1 + y_1}{s}$$

(or substituting $50b_w s/f_y$ for $2A_t$ whichever controls).

$$\frac{50b_w s}{f_y} = 0.0583 < 2A_t = 2 \times 0.0180 \times 5 = 0.18 \text{ in.}^2$$

Use $2A_t = 0.18 \text{ in.}^2$. Thus

$$A_l = \left(\frac{400 \times 14 \times 5}{60,000} \frac{500,000}{500,000 + \frac{15,000}{3 \times 0.0596}} - 0.18 \right) \frac{10.5 + 21.5}{5}$$

$$= 1.41 \text{ in.}^2 > 1.152 \text{ in.}^2$$

Therefore use $A_l = 1.41 \text{ in.}^2$.

Distribution of torsion longitudinal bars

Torsional $A_l = 1.41 \text{ in.}^2$. Assume that $\frac{1}{4}A_l$ goes to the top corners and $\frac{1}{4}A_l$ goes to the bottom corners of the stirrups, to be added to the flexural bars. The balance, $\frac{1}{2}A_l$, would thus be distributed equally to the vertical faces of the beam cross section at a spacing not to exceed 12 in. center to center.

$$\text{Midspan } \Sigma A_s = \frac{A_l}{4} + A_s = \frac{1.41}{4} + 3.4 = 3.75 \text{ in.}^2$$

Provide five No. 8 (25.4 mm diameter) bars at the bottom. Provide two No. 4 (12.7 mm diameter) bars with an area of 0.40 in.^{2} at the top. The required area of $A_l/4$ is 0.35 in.^{2}. The area of steel needed for each vertical face = 0.35 in.^{2}. Provide two No. 4 (12.7 mm diameter) bars on each side. Figure 5.15 shows the geometry of the cross section.

Solution

(b) Compatibility torsion:

Factored torsional moment (Step 1)

Given $T_u = 75,000$ in.-lb (8.47 kN/m). Using the results of case (a), one gets

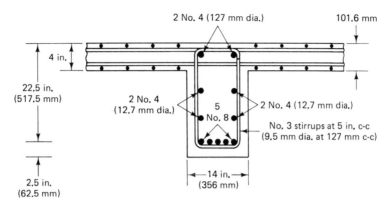

Figure 5.15 Web reinforcement details, Ex. 5.2(a).

$$\phi(0.5\,\sqrt{f_c'}\,\Sigma\,x^2y) = 0.85 \times 0.5 \times \sqrt{4000} \times 5284 \text{ in.}^3$$
$$= 142{,}030.5 \text{ in.-lb (16.04 kN-m)} > T_u$$
$$= 75{,}000 \text{ in.-lb}$$

So torsion effects can be neglected and only design for shear is needed.

Solution

(c) Compatibility torsion:

Factored torsional moment (Step 1)

Given that T_u = 300,000 in.-lb (33.9 kN-m) is greater than $\phi(0.5\sqrt{f_c'}\,\Sigma\,x^2y)$. So stirrups should be provided. Because this is a compatibility torsion, the section can be designed for a torsional moment of $\phi(4\sqrt{f_c'}\,\Sigma\,x^2y/3)$ if the external torsion exceeds this value.

$$\phi\!\left(4\sqrt{f_c'}\,\frac{\Sigma\,x^2y}{3}\right) = 378{,}748 \text{ in.-lb} > \text{given } T_u = 300{,}000 \text{ in.-lb}$$

Thus the section should be designed for T_u = 300,000 in.-lb.

Torsion closed stirrup design (Step 2)

Using Eq. 5.20b yields

$$T_c = 253{,}467 \text{ in.-lb (28.64 kN-m)}$$

$$T_s = T_n - T_c = \frac{300{,}000}{0.85} - 253{,}467 = 99{,}474 \text{ in.-lb (11.24 kN-m)}$$

$$\frac{A_t}{s} = \frac{99{,}474}{60{,}000 \times 1.34 \times 10.5 \times 21.5} = 0.0055 \text{ in.}^2/\text{in.}/\text{one leg.}$$

Shear stirrup design (Step 3)

$$V_c = \frac{2\sqrt{4000} \times 14 \times 22.5}{\sqrt{1 + (2.5 \times 0.0596 \times 300{,}000/15{,}000)^2}} = 12{,}676 \text{ lb } (57.04 \text{ kN})$$

$$V_s = V_n - V_c = \frac{15{,}000}{0.85} - 12{,}676 = 4971 \text{ lb } (22.40 \text{ kN})$$

$$\frac{A_v}{s} = \frac{4971}{60{,}000 \times 22.5} = 0.0037 \text{ in.}^2/\text{in.}/\text{two legs}$$

Combined closed stirrups for torsion and shear (Step 4)

$$\frac{A_{vt}}{s} = \frac{2A_t}{s} + \frac{A_v}{s} = 2 \times 0.0055 + 0.0037 = 0.0147 \text{ in.}^2/\text{in.}/\text{two legs}$$

Trying No. 3 ties with an area $= 2 \times 0.11 = 0.22$ in.^{2} (9.5 mm diameter, $A_s = 142$ mm^2) yields

$$s = \frac{\text{area of tie cross section } A_s}{\text{required } A_{vt}/s} = \frac{0.22}{0.0147} = 14.96 \text{ in.}$$

$$\text{Maximum permissible spacing } s_{max} = \frac{x_1 + y_1}{4} = \frac{10.5 + 21.5}{4} = 8 \text{ in.} < 14.96 \text{ in.}$$

So provide No. 3 (9.5 mm diameter) closed stirrups at 8 in. center to center (203.2 mm center to center).

$$\text{Minimum stirrup area required} = \frac{50 \times 14 \times 8}{60{,}000} = 0.0933 \text{ in.}^2$$

$$\text{Area provided} = 0.22 \text{ in.}^2 > 0.0933 \text{ in.}^2 \quad \text{O.K.}$$

Longitudinal torsion steel design (Step 5)

$$A_l = 2A_t \frac{x_1 + y_1}{s} = 2 \times 0.0055(10.5 + 21.5) = 0.32 \text{ in.}^2$$

$$\frac{50b_w s}{f_y} = 0.093 > 2A_t = 0.011 \times 8 = 0.088 \text{ in.}^2$$

Thus alternatively,

$$A_t = \left(\frac{400 \times 8 \times 14}{60{,}000} \frac{300{,}000}{300{,}000 + \frac{15{,}000}{3 \times 0.0596}} - 0.093 \right) \frac{10.5 + 21.5}{8}$$

$$= 1.96 \text{ in.}^2$$

Therefore A_t to be provided $= 1.96$ in.^{2}.

Distribution of torsion longitudinal bars

Torsional $A_t = 1.96$ in.^{2}, so $A_t/4 = 0.49$ in.^{2}. Using the same logic as that followed in case (a), provide five No. 8 (25.4 mm diameter) bars at the bottom face. The area required, $A_s + A_t/4 = 3.89$ in.^{2}; the area provided $= 3.95$ in.^{2}. The required area at the top corners and at each vertical face $= A_t/4 = 0.49$ in.^{2}. Provide two No. 5 bars (15.9 mm diameter) at the top and at each of two vertical sides, giving 0.62 in.^{2} in each area. Figure 5.16 shows the geometry of the section reinforcement.

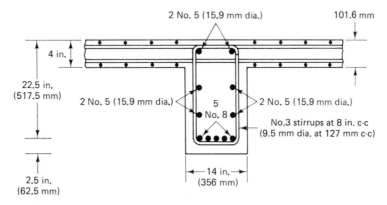

Figure 5.16   Web reinforcement details, Ex. 5.2(c).

SELECTED REFERENCES

5.1 Hsu, T. T. C., *Torsion of Reinforced Concrete*, Van Nostrand Reinhold, New York, 1983, 510 pp.

5.2 Mattock, A. H., "Diagonal Tension Cracking in Concrete Beams with Axial Forces," *Journal of the Structural Division, ASCE*, Vol. 95, No. ST9, September 1969, pp. 1887–1900.

5.3 Park, R., and T. Paulay, *Reinforced Concrete Structures*, Wiley, New York, 1975, 768 pp.

5.4 Rangan, B. V., and A. J. Hall, "Strength of Rectangular Prestressed Concrete Beams in Combined Torsion, Bending and Shear," *Journal of the American Concrete Institute*, Proc. Vol. 70, April 1973, pp. 270–279.

5.5 Taylor, H. P. J., "The Fundamental Behavior of Reinforced Concrete Beams in Bending and Shear," *Special Publication SP42*, Vol. 1, American Concrete Institute, Detroit, 1974, pp. 43–77.

5.6 Zia, P., "Tension Theories for Concrete Members," *Special Publication SP 18-4*, American Concrete Institute, Detroit, 1968, pp. 103–132.

5.7 Zsutty, T. C., "Beam Shear Strength Prediction by Analysis of Existing Data," *Journal of the American Concrete Institute*, Proc. Vol. 65, November 1968, pp. 943–951.

5.8 ACI Committee 340, *Strength Design Handbook,* Vol. 1: *Beams, Slabs, Brackets, Footings and Pile Caps,* Special Publication SP17(84), American Concrete Institute, Detroit, 1984, 384 pp.

5.9 ACI-ASCE Committee 426, "Suggested Revisions to Shear Provisions for Building Codes," ACI 426 IR77, American Concrete Institute, Detroit, 1979, 84 pp.

PROBLEMS FOR SOLUTION

5.1 A simply supported beam has a clear span $l_n = 22$ ft (6.70 m) and is subjected to an external uniform service dead load $w_D = 1200$ lb per ft (17.5 kN/m) and live load $w_L = 900$ lb per ft (13.1 kN/m). Determine the maximum factored vertical shear V_u at the critical section. Also determine the nominal shear resistance V_c by both the short method and by the more refined method of taking the contribution of the flexural steel into account. Design the size and spacing of the diagonal tension reinforcement. Given:

$$b_w = 12 \text{ in. (304.8 mm)}$$
$$d = 17 \text{ in. (431.8 mm)}$$
$$h = 20 \text{ in. (508.0 mm)}$$
$$A_s = 6.0 \text{ in.}^2 \text{ (3780 mm}^2)$$
$$f'_c = 4000 \text{ psi (27.6 MPa), normalweight concrete}$$
$$f_y = 60,000 \text{ psi (413.7 MPa)}$$

Assume that no torsion exists.

5.2 Solve Problem 6.1, assuming that the beam is made of sand-lightweight concrete and that it is subjected to an axial service compressive load of 2500 lb acting at its plastic centroid.

5.3 Design the vertical stirrups for a beam having the shear diagram shown in Fig. 5.17, assuming that $V_c = 2\sqrt{f'_c}b_w d$. Given:

$$b_w = 14 \text{ in. (355.6 mm)}$$
$$d_w = 20 \text{ in.}$$
$$V_{u1} = 75,000 \text{ lb (333.6 kN)}$$
$$V_{u2} = 60,000 \text{ lb (266.9 kN)}$$

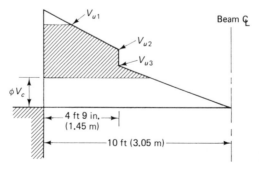

Figure 5.17

5.4 Calculate the torsional capacity T_c for the sections shown in Fig. 5.18. Given:

$V_u/T_u = 0.05$

$f'_c = 4000$ psi (27.6 MPa), normalweight concrete

$f_y = 60,000$ psi (413.7 MPa)

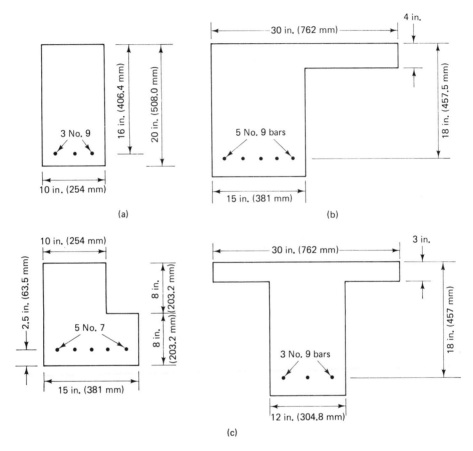

Figure 5.18   Cross sections for Problem 5.4.

5.5 A cantilever beam is subjected to a concentrated service live load of 20,000 lb (90 kN) acting at a distance of 3 ft 6 in. (1.07 m) from the wall support. In addition, the beam has to resist an equilibrium factored torsion $T_u = 300,000$ in.-lb (33.89 kN/m). The beam cross section is 12 in. × 24 in. (304.8 mm × 609.6 mm) with an effective depth of 22.5 in. (571.5 mm). Design the stirrups and the additional longitudinal steel needed. Given:

$f'_c = 3500$ psi

$f_y = 60,000$ psi

$A_s = 4.0$ in.^{2} (2580.64 mm^2)

5.6 The first interior span of a four-span continuous beam has a clear span $l_n = 18$ ft (5.49 m). The beam is subjected to a uniform external service dead load $w_D = 1700$ plf (24 pkN/m) and a service live load $w_L = 2200$ plf (32.1 kN/m). Design the section for flexure, diagonal tension, and torsion. Select the size and spacing of the closed stirrups and extra longitudinal steel that might be needed for torsion. Assume that the beam width $b_w = 15$ in. (381.0 mm) and that redistribution of torsional stresses is possible such that the external torque T_u can be assumed as $\phi(4\sqrt{f_c'} \, \Sigma \, x^2 y / 3)$. Given:

$f_c' = 5000$ psi (34.47 MPa), normalweight concrete

$f_y = 60,000$ psi (413.7 MPa)

CHAPTER

6

Deflection and Cracking

6.1 SIGNIFICANCE OF DEFLECTION OBSERVATION

The working stress method of design and analysis used prior to the 1970s limited the stress in concrete to about 45% of its compressive strength and the stress in the steel to less than 50% of its yield strength. Elastic analysis was applied to the design of structural frames as well as reinforced concrete sections. Structural elements were proportioned to carry the highest service-level moment along the span of the member, with redistribution of moment effect often largely neglected. As a result, heavier sections with higher reserve strength resulted compared to those obtained by the current ultimate strength approach.

Higher-strength concretes having f'_c values in excess of 12,000 psi (82.74MPa) and higher-strength steels are being used in strength design and expanding knowledge of the properties of the materials has resulted in lower values of load factors and reduced reserve strength. Thus more slender and efficient members are specified with deflection becoming a more pronounced controlling criterion.

Beams and slabs are rarely built as isolated members but are a monolithic part of an integrated system. Excessive deflection of a floor slab may cause dislocations in the partitions that it supports. Excessive deflection of a beam can damage a partition below and excessive deflection of a lintel beam above a window opening could crack the glass panels. In the case of open floors or roofs, such as top garage floors, ponding of water can result. For these reasons, deflection control criteria are necessary for proper design.

6.2 DEFLECTION BEHAVIOR OF BEAMS

The load–deflection relationship of a reinforced concrete beam is basically trilinear, as idealized in Fig. 6.1. It is composed of three regions prior to rupture:

Region I: Precracking stage where a structural member is crack-free (Fig. 6.2)

Region II: Postcracking stage where the structural member develops acceptable controlled cracking both in distribution and width

Region III: Postserviceability cracking stage where the stress in the tension reinforcement reaches the limit state of yielding

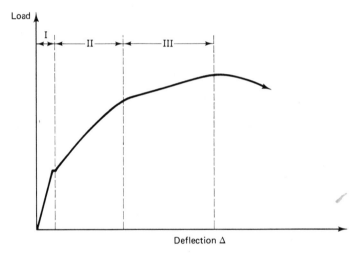

Figure 6.1 Beam load-deflection relationship. Region I, precracking stage; region II, postcracking stage; region III, postserviceability stage (steel yields).

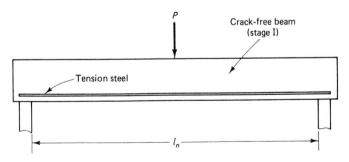

Figure 6.2 Centrally loaded beam at the precracking stage.

6.2.1 Precracking Stage: Region I

The precracking segment of the load–deflection curve is essentially a straight line defining full elastic behavior. The maximum tensile stress in the beam in this region is less than its tensile strength in flexure—that is, less than the modulus of rupture f_r of concrete. The flexural stiffness EI of the beam can be estimated by using Young's modulus E_c of concrete and the moment of inertia of the uncracked reinforced concrete cross section. The load–deflection behavior depends on the stress–strain relationship of the concrete as a significant factor. A typical stress–strain diagram of concrete is shown in Fig. 6.3.

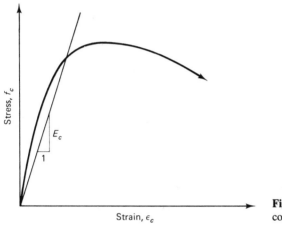

Stress, f_c

E_c

1

Strain, ϵ_c

Figure 6.3   Stress-strain diagram of concrete.

The value of E_c can be estimated by using the ACI empirical expression given in Chapter 2:

$$E_c = 33w^{1.5}\sqrt{f_c'}$$

or $\quad E_c = 57{,}000\sqrt{f_c'} \quad$ for normalweight concrete

An accurate estimation of the moment of inertia I means considering the contribution of the steel reinforcement A_s. It can be done by replacing the steel area by an equivalent concrete area $(E_s/E_c)A_s$, for the value of Young's modulus E_s of the reinforcement is higher than E_c. One can transform the steel area to an equivalent concrete area, calculate the center of gravity of the transformed section, and obtain the transformed moment of inertia I_{gt}.

Example 6.1 presents a typical calculation of I_{gt} for a transformed rectangular section. Most designers, however, use a gross moment of inertia I_g based on the uncracked concrete section, disregarding the additional stiffness contributed by the steel reinforcement as insignificant.

The precracking region stops at the initiation of the first flexural crack when the concrete stress reaches its modulus of rupture strength f_r. Similarly to the direct tensile splitting strength, the modulus of rupture of concrete is proportional to the square root of its compressive strength. For design purposes, the value of the modulus for normal weight concrete may be taken as

$$f_r = 7.5\sqrt{f'_c} \tag{6.1}$$

If lightweight concrete is used, the value of f_r from Eq. 6.1 is multiplied by 0.75 for all lightweight concrete and by 0.85 for sand-lightweight concrete.

If the distance of the extreme tension fiber from the center of gravity of the section is y_t and the cracking moment is M_{cr}, then

$$M_{cr} = \frac{I_g f_r}{y_t} \tag{6.2}$$

For a rectangular section,

$$y_t = \frac{h}{2} \tag{6.3}$$

where h is the total thickness of the beam. Equation 6.2 is derived from the classical bending equation $\sigma = Mc/I$ for elastic and homogeneous materials.

Calculations of deflection for this region are not important, for few reinforced concrete beams remain uncracked under actual loading. Mathematical knowledge of the variation in stiffness properties is important, because segments of the beam along the span in the actual structure can remain uncracked.

6.2.1.1 Example 6.1: Alternative Methods of Cracking Moment Evaluation

Calculate the cracking moment M_{cr} for the beam cross section shown in Fig. 6.4, using both (a) transformed and (b) gross cross-section alternatives in the solution. Given:

$f'_c = 4000$ psi (27.6 MPa)

$f_y = 60,000$ psi (414 MPa)

$E_s = 29 \times 10^6$ psi (200,000 MPa), normalweight concrete

Reinforcement: four No. 9 bars (four bars 28.6 mm diameter) placed in two bundles.

Solution

(a) *Transformed section solution:* Depth of center-of-gravity axis, $\bar{y}$, can be obtained by using the first moment of area—namely,

$$\left[bh + \left(\frac{E_s}{E_c} - 1\right)A_s \right]\bar{y} = bh\frac{h}{2} + \left(\frac{E_s}{E_c} - 1\right)A_s d$$

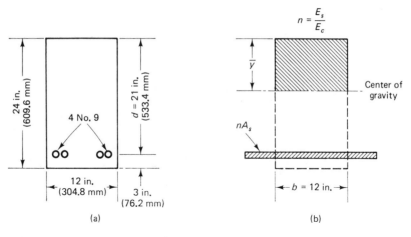

Figure 6.4   Cross-section transformation in Ex. 6.1: (a) midspan section; (b) transformed section.

Note that $(E_s/E_c) - 1$ is used instead of E_s/E_c to account for the concrete displaced by the reinforcing bars.

It is customary to denote $n = E_s/E_c$ as the modular ratio. Taking moments about the top extreme fibers of the section,

$$\bar{y} = \frac{(bh^2/2) + (n - 1)A_s d}{bh + (n - 1)A_s}$$

For normalweight 4000-psi concrete,

$$E_c = 57,000\sqrt{4000}$$

$$= 3.6 \times 10^6 \text{ psi } (24.8 \times 10^6 \text{ MPa})$$

$$n = \frac{29 \times 10^6}{3.6 \times 10^6} = 8.1$$

$$\bar{y} = \frac{\dfrac{12 \times (24)^2}{2} + (8.1 - 1)4.0 \times 21}{12 \times 24 + (8.1 - 1)4.0} = 12.8 \text{ in. } (325.1 \text{ mm})$$

If the moment of inertia of steel reinforcement about its own axis is neglected as insignificant, then

$$\text{Transformed section } I_{gt} = \frac{bh^3}{12} + bh(12.8 - 12.0)^2 + (n - 1)A_s(d - \bar{y})^2$$

or

$$I_{gt} = \frac{12 \times 24^3}{12} + 12 \times 24 \times 0.8^2 + 7.1 \times 4.0(21 - 12.8)^2$$

$$= 15,918 \text{ in.}^4 (66.22 \times 10^8 \text{ mm}^4)$$

Photo 45 Ladd Canyon Overpass, Oregon. (Courtesy of Portland Cement Association.)

The distance of the center of gravity of the transformed section from the lower extreme fibers is

$$y_t = 24 - 12.8 = 11.2 \text{ in. } (284.5 \text{ mm})$$

$$f_r = 7.5\sqrt{4000} = 474.3 \text{ psi } (3.27 \text{ MPa})$$

$$M_{cr} = \frac{I_g f_r}{y_t} = \frac{15{,}918 \times 474.3}{11.2} = 674{,}100 \text{ in.-lb } (76.17 \text{ kN-m})$$

(b) *Gross cross-section solution:*

$$\bar{y} = \frac{h}{2} = 12 \text{ in.}$$

$$\text{Cross section } I_g = \frac{bh^3}{12} = \frac{12 \times 24^3}{12} = 13{,}824 \text{ in.}^4$$

$$y_t = 12 \text{ in. } (304.8 \text{ mm})$$

$$f_r = 474.3 \text{ psi}$$

$$M_{cr} = \frac{13,824 \times 474.3}{12} = 546,394 \text{ in.-lb } (61.74 \text{ kN-m})$$

There is a difference of about 15% in the value of I_g and 19% in the value of M_{cr}. Even though this percentage difference in the values of the I_g and M_{cr} obtained by the two methods seems somewhat high, such a difference in the deflection calculation values is not significant and, in most cases, does not justify using the transformed-section method for evaluating M_{cr}.

6.2.2 Postcracking Service Load Stage: Region II

The precracking region ends at the initiation of the first crack and moves into region II of the load–deflection diagram in Fig. 6.1. Most beams lie in this region at service loads. A beam undergoes varying degrees of cracking along the span corresponding to the stress and deflection levels at each section. So cracks are wider and deeper at midspan whereas only narrow minor cracks develop near the supports in a simple beam.

When flexural cracking develops, the contribution of the concrete in the tension zone reduces substantially. Thus the flexural rigidity of the section is reduced, making the load–deflection curve less steep in this region than in the precracking stage segment. As the magnitude of cracking increases, stiffness continues to decrease, reaching a lower-bound value corresponding to the reduced moment of inertia of the cracked section. At this limit state of service load cracking, the contribution of tension-zone concrete to the stiffness is neglected. The moment of inertia of the cracked section designated as I_{cr} can be calculated from the basic principles of mechanics.

Strain and stress distribution across the depth of a typical cracked rectangular concrete section are shown in Fig. 6.5. The following assumptions are made with respect to deflection computation based on extensive testing verification:

1. The strain distribution across the depth is assumed to be linear.
2. Concrete does not resist any tension.
3. Both concrete and steel are within the elastic limit.
4. Strain distribution is similar to that assumed for strength design, but the magnitudes of strains, stresses, and stress distribution are different.

To calculate the moment of inertia, the value of the neutral axis depth, c, should be determined from horizontal force equilibrium.

$$A_s f_s = bc \frac{f_c}{2} \tag{6.4a}$$

Because steel stress $f_s = E_s \epsilon_s$ and concrete stress, $f_c = E_c \epsilon_c$, Eq. 6.4a can be rewritten

$$A_s E_s \epsilon_s = \frac{bc}{2} E_c \epsilon_c \tag{6.4b}$$

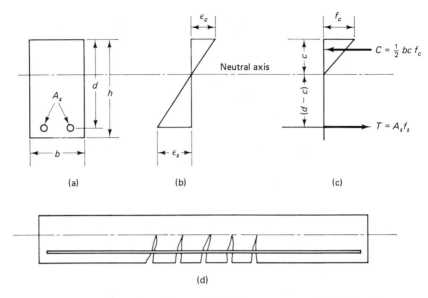

Figure 6.5 Elastic strain and stress distributions across a cracked reinforced concrete section: (a) cross section; (b) strain; (c) elastic stress and force; (d) cracked beam prior to failure in flexure.

From similar triangles in Fig. 6.5b

$$\frac{\epsilon_c}{c} = \frac{\epsilon_s}{d - c} \tag{6.5a}$$

or

$$\epsilon_s = \epsilon_c\left(\frac{d}{c} - 1\right) \tag{6.5b}$$

From Eqs. 6.4b and 6.5b

$$A_s E_s \epsilon_c\left(\frac{d}{c} - 1\right) = \frac{bc}{2} E_c \epsilon_c \tag{6.6a}$$

or

$$\frac{A_s E_s}{E_c}\left(\frac{d}{c} - 1\right) = \frac{bc}{2} \tag{6.6b}$$

By replacing the modular ratio E_s/E_c by n, Eq. 6.6b can be rewritten

$$\frac{bc^2}{2} + nA_s c - nA_s d = 0 \tag{6.6c}$$

The value of c can be obtained by solving the quadratic equation 6.6c. The moment of inertia I_{cr} can be obtained from

$$I_{cr} = \frac{bc^3}{3} + nA_s(d - c)^2 \tag{6.7}$$

where the term $bc^3/3$ in Eq. 6.7 denotes the moment of inertia of the *compressive* area bc about the neutral axis—that is, the base of the compression rectangle, neglecting the section area in tension below the neutral axis. The reinforcing area is multiplied by n to transform it to its equivalent in concrete for contribution to the section stiffness. The moment of inertia of the steel about its own axis is disregarded as negligible.

Only part of the beam cross section is cracked in the case under discussion. As seen from Fig. 6.5d, the uncracked segments below the neutral axis along the beam span possess some degree of stiffness that contributes to the overall beam rigidity. The actual stiffness of the beam lies between $E_c I_g$ and $E_c I_{cr}$, depending on such other factors as (a) extent of cracking, (b) distribution of loading, and (c) contribution of the concrete, as seen in Fig. 6.5d between the cracks. Generally as the load approaches the steel yield load level, the stiffness value approaches $E_c I_{cr}$.

Branson developed simplified expressions for calculating the effective stiffness $E_c I_e$ for design. The Branson equation, verified as applicable to most cases of reinforced and prestressed beams and universally adopted for deflection calculations, defines the effective moment of inertia as

$$I_e = \left(\frac{M_{cr}}{M_a}\right)^3 I_g + \left[1 - \left(\frac{M_{cr}}{M_a}\right)^3\right] I_{cr} \leq I_g \qquad (6.8a)$$

Equation 6.8a is also written in the form

$$I_e = I_{cr} + \left(\frac{M_{cr}}{M_a}\right)^3 (I_g - I_{cr}) \leq I_g \qquad (6.8b)$$

The effective moment of inertia I_e as shown in Eq. 6.8b depends on the maximum moment M_a along the span in relation to the cracking moment capacity M_{cr} of the section.

6.2.2.1 Example 6.2: Effective Moment of Inertia of Cracked Beam Sections

Calculate the moment of inertia I_{cr} and the effective moment of inertia I_e of the beam cross section in Ex. 6.1 if the external maximum service load moment is 2,000,000 in.-lb (226 kN-m). Given (Ex. 6.1):

$b = 12$ in.

$d = 21$ in.

$h = 24$ in.

$A_s = 4.0$ in.^{2}

$f'_c = 4000$ psi

$f_y = 60,000$ psi

$E_s = 29 \times 10^6$ psi

$E_c = 3.6 \times 10^6$ psi

$n = 8.1$

Solution

From Eq. 6.6c

$$\frac{12c^2}{2} + 8.1 \times 4.0c - 8.1 \times 4.0 \times 21 = 0$$

So neutral axis depth $c = 8.3$ in. (210.8 mm). From Eq. 6.7

$$I_{cr} = \frac{12.0 \times 8.3^3}{3} + 8.1 \times 4.0(21.0 - 8.3)^2 = 7513 \text{ in.}^4(31.25 \times 10^8 \text{ mm}^4)$$

Using the I_{gt} and M_{cr} values of Ex. 6.1, which include the effect of the transformed steel area,

$$I_e = 7513 + \left(\frac{674,100}{2,000,000}\right)^3 (15,918 - 7513)$$

$$= 7835 \text{ in.}^4(32.59 \times 10^8 \text{ mm}^4) < I_g \qquad \text{as expected}$$

If the gross cross-section values for I_g and M_{cr} are used without including the effect of transformed A_s, the effective moment of inertia becomes

$$I_e = 7413 + \left(\frac{546,394}{2,000,000}\right)^3 (13,824 - 7513)$$

$$= 7462 \text{ in.}^4(31.79 \times 10^8 \text{ mm}^4) < I_g$$

Comparison of the two values of effective I_e calculated by the two methods (7835 in.^{4} versus 7642 in.^{4}) shows an insignificant difference. Thus use of the gross section properties in Eq. 6.8 is, in most cases, adequate particularly when one considers the variability in the loads and the randomness in the properties of concrete.

6.3 LONG-TERM DEFLECTION

Time-dependent factors magnify the magnitude of deflection with time. Consequently, the design engineer must evaluate immediate as well as long-term deflection in order to ensure that their values satisfy the maximum permissible criteria for the particular structure and its particular use.

Time-dependent effects are caused by the superimposed creep, shrinkage, and temperature strains. These additional strains induce a change in the distribution of stresses in the concrete and the steel, resulting in an increase in the curvature of the structural element for the same external load.

The calculation of creep and shrinkage strains at a given time is a complex process. One has to consider how these time-dependent concrete strains affect the stress in the steel and the curvature of the concrete element. In addition, the effect of progressive cracking on the change in stiffness factors must be considered, which considerably complicates the analysis and design process. Consequently, an empirical approach to evaluate deflection under sustained loading is, in many cases, more practical.

Photo 46   Deflected simply supported beam prior to failure. (Tests by Nawy, et al.)

Photo 47   Deflected continuous prestressed beam prior to failure. (Tests by Nawy, Potyondy, et al.)

The additional deflection under sustained loading and long-term shrinkage in accordance with the ACI procedure can be calculated by using a multiplying factor:

$$\lambda = \frac{T}{1 + 50\rho'} \tag{6.9}$$

where ρ' is the compression reinforcement ratio calculated at midspan for simple and continuous beams and T is a factor that is taken as 1.0 for loading time duration of 3 months, 1.2 for 6 months, 1.4 for 12 months, and 2.0 for 5 years or more.

If the instantaneous deflection is Δ_i, the additional time-dependent deflection becomes $\lambda\Delta_i$ and the total long-term deflection would be $(1 + \lambda)\Delta_i$. Because live loads are not present at all times, only part of the live load, plus the more permanent dead load, is considered as the sustained load. Figure 6.6 gives the relationship between the load duration in months and the multiplier T in Eq. 6.6. It is seen from this plot that the maximum multiplier value $T = 2.0$ represents a nominal limiting time-dependent factor for 5 years' duration of loading. In effect, the expression for the long-term factor λ in Eq. 6.9 has similar characteristics as the stiffness EI of a section in that it is a function of the material property T and the section property $(1 + 50\rho')$.

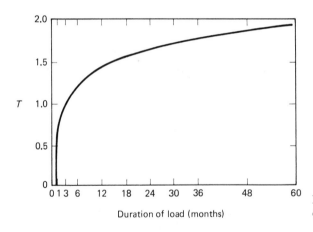

Figure 6.6 Multipliers for long-term deflection.

The total long-term deflection is

$$\Delta_{LT} = \Delta_L + \lambda_\infty \Delta_D + \lambda_t \Delta_{LS} \tag{6.10}$$

where Δ_L = immediate live load deflection
  Δ_D = immediate dead load deflection
  Δ_{LS} = sustained live load deflection (a percentage of the immediate Δ_L determined by expected duration of sustained load)
  λ_∞ = time-dependent multiplier for infinite duration of sustained load
  λ_t = time-dependent multiplier for limited load duration

The value of the multiplier λ is the same for normal or lightweight concrete.

6.4 PERMISSIBLE DEFLECTIONS IN BEAMS AND ONE-WAY SLABS

Permissible deflections in a structural system are governed primarily by the amount that can be sustained by the interacting components of a structure without loss of aesthetic appearance and without detriment to the deflecting member. The level of acceptability of deflection values is a function of such factors as the type of building, the use or nonuse of partitions, the presence of plastered ceilings, or the sensitivity of equipment or vehicular systems that are being supported by the floor. Because deflection limitations must be placed at service load levels, structures designed conservatively for low concrete and steel stresses would normally have no deflection problems. Present-day structures, however, are designed by ultimate load procedures efficiently using high-strength concretes and steels. More slender members resulting from such designs would need to be better controlled for serviceability deflection performance, both immediate and long term.

6.4.1 Empirical Method of Minimum Thickness Evaluation for Deflection Control

The ACI code recommends (Table 6.1) minimum thickness for beams as a function of the span length, where no deflection computations are necessary if the member is not supporting or attached to construction likely to be damaged by large deflections.

TABLE 6.1 MINIMUM THICKNESS OF BEAMS AND SLABS UNLESS DEFLECTIONS ARE COMPUTED[a]

Member[b]	Minimum thickness, h			
	Simply supported	One end continuous	Both ends continuous	Cantilever
Solid one-way slabs	$l/20$	$l/24$	$l/28$	$l/10$
Beams or ribbed one-way slabs	$l/16$	$l/18.5$	$l/21$	$l/8$

[a]Clear span length l is in inches. Values given should be used directly for members with normalweight concrete ($w_c = 145$ pcf) and grade 60 reinforcement. For other conditions, the values should be modified as follows:

1. For structural lightweight concrete having unit weights in the range 90 to 120 lb/ft³, the values should be multiplied by $(1.65 - 0.005w_c)$ but not less than 1.09, where w_c is the unit weight in lb/ft³.
2. For f_y other than 60,000 psi, the values should be multiplied by $(0.4 + f_y/100,000)$.

[b]Members not supporting or attached to partitions or other construction likely to be damaged by large deflections.

TABLE 6.2 MINIMUM PERMISSIBLE RATIOS OF SPAN (ℓ) TO DEFLECTION (Δ)
(ℓ = *longer span*)

Type of member	Deflection, Δ, to be considered	$(l/\Delta)_{min}$
Flat roofs not supporting and not attached to nonstructural elements likely to be damaged by large deflections	Immediate deflection due to live load L	180[a]
Floors not supporting and not attached to nonstructural elements likely to be damaged by large deflections	Immediate deflection due to live load L	360
Roof or floor construction supporting or attached to nonstructural elements likely to be damaged by large deflections	That part of total deflection occurring after attachment of nonstructural elements; sum of long-term deflection due to all sustained loads (dead load plus any sustained portion of live load) and immediate deflection due to any additional live load[b]	480[c]
Roof or floor construction supporting or attached to nonstructural elements not likely to be damaged by large deflections		240[c]

[a]Limit not intended to safeguard against ponding. Ponding should be checked by suitable calculations of deflection, including added deflections due to ponded water, and considering long-term effects of all sustained loads, camber, construction tolerances, and reliability of provisions for drainage.

[b]Long-term deflection has to be determined but may be reduced by the amount of deflection calculated to occur before attachment of nonstructural elements. This reduction is made on the basis of accepted engineering data relating to time–deflection characteristics of members similar to those being considered.

[c]Ratio limit may be lower if adequate measures are taken to prevent damage to supported or attached elements, but should not be lower than tolerance of nonstructural elements.

Other deflections would need to be calculated and controlled as in Table 6.2. If the total beam thickness is less than required by the table, the designer should verify the deflection serviceability performance of the beam through detailed computations of the immediate and long-term deflections.

6.4.2 Permissible Limits of Calculated Deflection

The ACI code requires that the calculated deflection for a beam or one-way slab satisfy the serviceability requirement of minimum permissible deflection for the various structural conditions listed in Table 6.2 of Section 6.4.1 if Table 6.1 is *not* used. Long-term effects, however, cause measurable increases in deflection with time and sometimes result in excessive overstress in the steel and concrete. So it is always advisable to calculate the total time-dependent deflection Δ_{LT} in Eq. 6.10 and design the beam size based on the permissible span/deflection ratios of Table 6.2.

6.5 COMPUTATION OF DEFLECTIONS

6.5.1 Operational Deflection Calculation Procedure and Flowchart

Deflection of structures affects their aesthetic appearance as well as their long-term serviceability. The following step-by-step procedure should be followed after the structural member is designed for flexure.

1. Compare the total design depth of the member with the minimum allowable value obtained from Table 6.1. If it is less than the allowable, proceed to perform a detailed calculation of short-term and long-term deflection. It is, however, always advisable to perform the detailed calculations regardless of the comparison with Table 6.1.

2. The detailed calculations should establish as a first step:
    (a) The gross moment of inertia I_g
    (b) The cracking moment M_{cr}, which is a function of the modulus of rupture of concrete

3. Calculate the depth c of the neutral axis of the *transformed* section. Find the cracking moment of inertia I_{cr}.

4. Find the effective moment of inertia I_e as follows:

$$I_e = \left(\frac{M_{cr}}{M_a}\right)^3 I_g + \left[1 - \left(\frac{M_{cr}}{M_a}\right)^3\right] I_{cr} \le I_g$$

or

$$I_e = I_{cr} + \left(\frac{M_{cr}}{M_a}\right)^3 (I_g - I_{cr}) \le I_g$$

The effective I_e should be calculated for the following service load level combinations:
    (a) Dead load (D)
    (b) Dead load + sustained portion of live load ($D + \alpha L$, where α is less than 1.0)
    (c) Dead load + live load ($D + L$)

5. Calculate the immediate deflection based on I_e of the three combinations in step 4, using the applicable elastic deflection expression in Table 6.3.

6. Calculate the long-term deflection, finding first the time-dependent multiplier $\lambda = T/(1 + 50\rho')$ from values in Fig. 6.6. The total long-term deflection is

$$\Delta_{LT} = \Delta_L = \lambda_\infty \, \Delta_D + \lambda_t \, \Delta_{LS}$$

7. If $\Delta_{LT} <$ maximum permissible Δ in Table 6.2, limit the use of the structure to particular loading types or conditions or enlarge the section. Figure 6.7 shows a flowchart of the operational sequence of deflection control checks that the designer engineer should use when deflection computations are necessary.

6.5.2 Deflection Calculation Flowchart

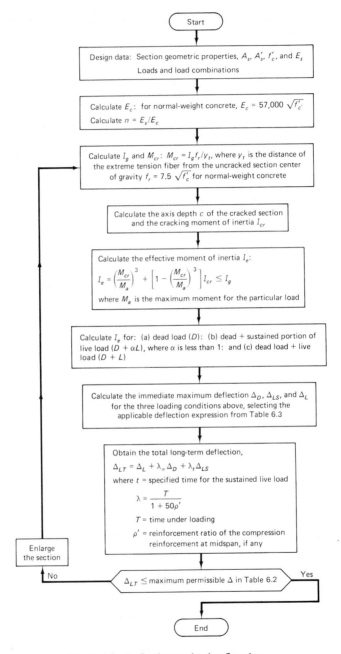

Figure 6.7   Deflection evaluation flowchart.

6.5.3 Governing Equations

The general expression for the maximum deflection Δ_{max} in an elastic member can be expressed from basic principles of mechanics as

$$\Delta_{max} = K\frac{Wl_n^3}{48EI_c}$$

where W = total load on the span
$\quad I_n$ = clear span length
$\quad E$ = modulus of concrete
$\quad I_c$ = moment of inertia of the section
$\quad K$ = a factor depending on the degree of fixity of the support

Equation 6.11 can also be written in terms of moment such that the deflection at any point in a beam is

$$\Delta = k\frac{ML^2}{E_cI_e}$$

where k = a factor depending on support fixity and load conditions
$\quad M$ = moment acting on the section
$\quad I_e$ = effective moment of inertia

Table 6.3 gives the maximum elastic deflection values in terms of the gravity load for typical beams loaded with uniform or concentrated load.

TABLE 6.3 MAXIMUM DEFLECTION EXPRESSIONS FOR MOST COMMON LOAD AND SUPPORT CONDITIONS

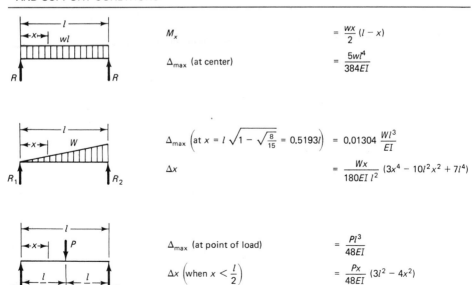

$M_x \qquad\qquad\qquad = \frac{wx}{2}(l-x)$

Δ_{max} (at center) $\qquad = \frac{5wl^4}{384EI}$

$\Delta_{max}\left(\text{at } x = l\sqrt{1-\sqrt{\frac{8}{15}}} = 0.5193l\right) = 0.01304\frac{Wl^3}{EI}$

$\Delta x \qquad\qquad\qquad = \frac{Wx}{180EI\,l^2}(3x^4 - 10l^2x^2 + 7l^4)$

Δ_{max} (at point of load) $\qquad = \frac{Pl^3}{48EI}$

$\Delta x \left(\text{when } x < \frac{l}{2}\right) \qquad = \frac{Px}{48EI}(3l^2 - 4x^2)$

TABLE 6.3 (Continued)

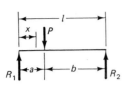

$$\Delta_{max}\left(\text{at } x = \sqrt{\frac{a(a+2b)}{3}} \text{ when } a > b\right) = \frac{Pab(a+2b)\sqrt{3a(a+2b)}}{27EI\,l}$$

$$\Delta a \text{ (at point of load)} = \frac{Pa^2 b^2}{3EI\,l}$$

$$\Delta x \text{ (when } x < a) = \frac{Pbx}{6EI\,l}(l^2 - b^2 - x^2)$$

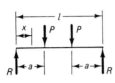

$$\Delta_{max} \text{ (at center)} = \frac{Pa}{24EI}(3l^2 - 4a^2)$$

$$\Delta x \text{ (when } x < a) = \frac{Px}{6EI}(3la - 3a^2 - x^2)$$

$$\Delta x \text{ (when } x > a \text{ and } < (l - a)) = \frac{Pa}{6EI}(3lx - 3x^2 - a^2)$$

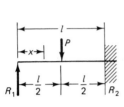

$$\Delta_{max}\left(\text{at } x = l\sqrt{\tfrac{1}{5}} = 0.4472l\right) = \frac{Pl^3}{48EI\sqrt{5}} = 0.009317\frac{Pl^3}{EI}$$

$$\Delta x \text{ (at point of load)} = \frac{7Pl^3}{768EI}$$

$$\Delta x \left(\text{when } x < \frac{l}{2}\right) = \frac{Px}{96EI}(3l^2 - 5x^2)$$

$$\Delta x \left(\text{when } x > \frac{l}{2}\right) = \frac{P}{96EI}(x - l)^2(11x - 2l)$$

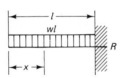

$$\Delta_{max} \text{ (at free end)} = \frac{wl^4}{8EI}$$

$$\Delta x = \frac{w}{24EI}(x^4 - 4l^3 x + 3l^4)$$

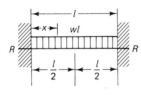

$$\Delta_{max} \text{ (at center)} = \frac{wl^4}{384EI}$$

$$\Delta x = \frac{wx^2}{24EI}(l - x)^2$$

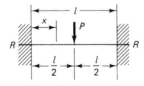

$$\Delta_{max} \text{ (at center)} = \frac{Pl^3}{192EI}$$

$$\Delta x \left(\text{when } x < \frac{l}{2}\right) = \frac{Px^2}{48EI}(3l - 4x)$$

TABLE 6.3   (Continued)

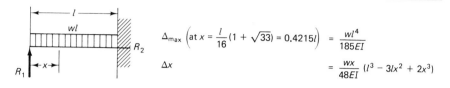

$$\Delta_{max}\left(at\ x = \frac{l}{16}(1 + \sqrt{33}) = 0.4215l\right) \quad = \quad \frac{wl^4}{185EI}$$

$$\Delta_x \qquad\qquad = \quad \frac{wx}{48EI}(l^3 - 3lx^2 + 2x^3)$$

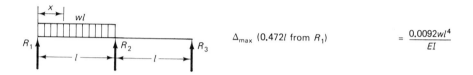

$$\Delta_{max}\ (0.472l\ \text{from}\ R_1) \qquad = \quad \frac{0.0092wl^4}{EI}$$

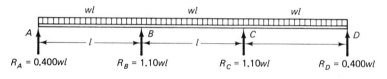

$$\Delta_{max}\ (0.446l\ \text{from}\ A\ \text{or}\ D) \quad = \quad \frac{0.0069wl^4}{EI}$$

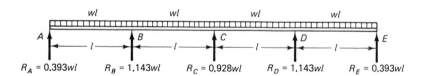

$$\Delta_{max}\ (0.440l\ \text{from}\ A\ \text{or}\ E) \quad = \quad \frac{0.0065wl^4}{EI}$$

6.5.4 Example 6.3: Deflection Behavior of a Uniformly Loaded Simple Span Beam

A simply supported uniformly loaded beam has a clear span $l_n = 27$ ft (8.23 m), a width $b = 10$ in. (254 mm), and a total depth $h = 16$ in. (406 mm), $d = 13.0$ in. (330 mm), and $A_s = 1.32$ in.^{2} (851.6 mm^2). It is subjected to a service dead load moment $M_D = 215,000$ in.-lb (24.3 kN-m), and a service live load moment $M_L = 250,000$ in.-lb

(28.3 kN-m). Determine if the beam satisfies the various deflection criteria for short-term and long-term loading. Assume that 60% of the live load is continuously applied for 24 months. Given:

$$f_c' = 5000 \text{ psi } (34.5 \text{ MPa}), \text{ normalweight concrete}$$

$$f_y = 60,000 \text{ psi } (413.7 \text{ MPa})$$

Solution

$$E_c = 33w^{1.5}\sqrt{5000} = 4.29 \times 10^6 \text{ psi } (29,580 \text{ MPa})$$

$$E_s = 29 \times 10^6 \text{ psi } (200,000 \text{ MPa})$$

$$\text{Modular ratio } n = \frac{E_s}{E_c} = \frac{29.0 \times 10^6}{4.29 \times 10^6} = 6.76$$

$$f_r = 7.5\sqrt{f_c'} = 7.5\sqrt{5000} = 530.3 \text{ psi } (3.66 \text{ MPa})$$

Minimum required depth

From Table 8.1

$$h_{min} = \frac{l_n}{16} = \frac{28.0 \times 12}{16} = 21.0 \text{ in. } (533.4 \text{ mm}) > \text{actual } h = 16.0 \text{ in.}$$

So deflection calculations must be made.

Effective moment of inertia I_e

$$I_g = \frac{bh^3}{12} = \frac{10(16)^3}{12} = 3413.3 \text{ in.}^4$$

$$y_t = \frac{16.0}{2} = 8.0 \text{ in.}$$

$$M_{cr} = \frac{f_r I_g}{y_t} = \frac{530.3 \times 3413.3}{8.0} = 226,259 \text{ in.-lb}$$

Depth of neutral axis c:

$$d = 16.0 - 3.0 = 13.0 \text{ in.} \qquad A_s = 1.32 \text{ in.}^2$$

$$\frac{10c^2}{2} = nA_s(d - c)$$

or $5c^2 = 6.76 \times 1.32(13.0 - c)$, to get $c = 4.03$ in.

$$I_{cr} = \frac{10c^3}{3} + 6.76 \times 1.32(13.0 - c)^2$$

$$= \frac{10(4.03)^3}{3} + 8.923(13.0 - 4.03)^2$$

$$= 936.1 \text{ in.}^4$$

Dead load:

$$\frac{M_{cr}}{M_a} = \frac{226{,}259}{215{,}000} = 1.05 > 1.0$$

Use $M_{cr} = M_a$ and $I_e = I_g$, for the dead load moment is smaller than the cracking moment (the beam will not crack at the dead load level).

Dead load + 60% live load:

$$\frac{M_{cr}}{M_a} = \frac{226{,}259}{215{,}000 + 0.6 \times 250{,}000} = 0.62$$

Dead load + live load:

$$\frac{M_{cr}}{M_a} = \frac{226{,}259}{215{,}000 + 250{,}000} = 0.49$$

$$I_e = \left(\frac{M_{cr}}{M_a}\right)^3 I_g + \left[1 - \left(\frac{M_{cr}}{M_a}\right)^3\right] I_{cr}$$

Dead load:

$$I_e = 3413.3 \text{ in.}^4$$

Dead load + 0.6 live load:

$$I_e = 0.24 \times 3413.3 + 0.76 \times 936.1 = 1530.6 \text{ in.}^4$$

Dead load + live load:

$$I_e = 0.12 \times 3413.3 + 0.88 \times 936.1 = 1233.4 \text{ in.}^4$$

Short-term deflection

$$\Delta = \frac{5wl^4}{384EI} = \frac{5Ml_n^2}{48EI} = \frac{5(27.0 \times 12)^2 M}{48 \times 4.29 \times 10^6 I} = 0.0025\frac{M}{I} \text{ in.}$$

Immediate live load deflection:

$$\Delta_L = \frac{0.0025(215{,}000 + 250{,}000)}{1233.4} - \frac{0.0025(215{,}000)}{3413.3}$$

$$= 0.943 - 0.158 = 0.785 \text{ in.}$$

Immediate dead load deflection:

$$\Delta_D = \frac{0.0025 \times 215{,}000}{3413.3} = 0.158 \text{ in.}$$

Immediate 60% live load deflection:

$$\Delta_{LS} = 0.0025\left(\frac{215{,}000 + 250{,}000 \times 0.6}{1530.6} - \frac{215{,}000}{3413.3}\right)$$

$$= 0.597 - 0.158 = 0.439 \text{ in.}$$

Long-term deflection

From Eq. 6.10

$$\Delta_{LT} = \Delta_L + \lambda_\infty \Delta_D + \lambda_t \Delta_{LS}$$

$$\lambda = \frac{T}{1 + 50\rho'} \qquad \text{where } \rho' = 0 \text{ for singly reinforced beam}$$

$$T \text{ for 5 years or more} = 2.0 \qquad \lambda_\infty = \frac{2.0}{1 + 0} = 2.0$$

$$T \text{ for 24 months} = 1.65 \qquad \lambda_t = \frac{1.65}{1} = 1.65$$

$$\Delta_{LT} = 0.785 + 2.0 \times 0.158 + 1.65 \times 0.439 = 1.825 \text{ in.}$$

Deflection requirements (Table 6.2)

$$\frac{l_n}{180} = \frac{27 \times 12}{180} = 1.80 \text{ in.} > \Delta_L$$

$$\frac{l_n}{360} = 0.90 \text{ in.} > \Delta_L$$

$$\frac{l_n}{480} = 0.68 \text{ in.} < \Delta_{LT}$$

$$\frac{l_n}{240} = 1.35 \text{ in.} < \Delta_{LT}$$

So the use of this beam is limited to floors or roofs not supporting or attached to nonstructural elements, such as partitions.

6.6 DEFLECTION CONTROL IN ONE-WAY SLABS

One-way slabs can be treated as rectangular beams of 12-in. (304.8-mm) width. Because floor loads are specified as load intensity per unit area, such intensity on a one-way slab over a 1-ft width becomes pounds per linear foot. Reinforcement is chosen in terms of bar spacing instead of number of bars and the area of steel for a 12-in. width of slab can easily be calculated for the total number of bars in a 12-in.-width strip.

6.6.1 Example 6.4: Deflection Calculations for a Simply Supported One-Way Slab

A 5-in.-thick ($h = 127$ mm) one-way slab has a span of 12 ft (3.66 in.). It is subjected to a live load of 60 psf (2.88 kPa) in addition to its self-weight. Calculate the immediate and long-term deflections of this slab, assuming that 45% of the live load is sustained over a 24-month period. Determine which type of elements it should support. Given:

$$f'_c = 3500 \text{ psi } (24.13 \text{ MPa})$$

$$f_y = 60,000 \text{ psi } (413.7 \text{ MPa})$$

$$E_s = 29 \times 10^6 \text{ psi } (200,000 \text{ MPa})$$

Steel reinforcement: No. 4 bars at 6 in. center-to-center spacing (12.7 mm diameter at 152.4 mm center to center)

Solution

Minimum depth requirement

From Table 6.1

$$\text{Minimum } h = \frac{l}{20} = \frac{12 \times 12}{20} = 7.20 \text{ in.}$$

$$\text{Actual } h = 5 \text{ in.} < 7.20 \text{ in.}$$

Deflection calculations must be made.

Material properties and bending moments

$$E_c = 57,000\sqrt{f'_c} = 57,000\sqrt{3500} = 3.37 \times 10^6 \text{ psi } (23,256 \text{ MPa})$$

$$E_s = 29 \times 10^6 \text{ psi } (200,000 \text{ MPa})$$

$$\text{Modular ratio } n = \frac{E_s}{E_c} = \frac{29 \times 10^6}{3.37 \times 10^6} = 8.61$$

Modulus of rupture $f_r = 7.5\sqrt{f'_c} = 7.5\sqrt{3500} = 443.7 \text{ psi}$

$$\text{Gross moment of inertia } I_g = \frac{bh^3}{12} = \frac{12(5.0)^3}{12} = 125.0 \text{ in.}^4$$

$$\text{Cracking moment } M_{cr} = \frac{f_r I_g}{y_t} = \frac{443.7 \times 125.0}{2.5} = 22,185 \text{ in.-lb}$$

$$\text{Service load bending moment} = \frac{wl_n^2}{8} = \frac{w(12.0)^2}{8} \times 12 \text{ in.-lb} = 216w \text{ in.-lb}$$

Neutral-axis depth of transformed section

$$A_s = \text{No. 4 at } 6 \text{ in.} = 0.40 \text{ in.}^2 \text{ per 12-in.-wide strip}$$

$$d = h - 0.75 - \frac{\phi}{2} = 5.0 - 0.75 - 0.25 = 4.0 \text{ in.}$$

If c is the depth from the compression fibers to the neutral axis of the transformed section, from Eq. 6.6c for rectangular sections

$$\frac{bc^2}{2} + nA_s c - nA_s d = 0$$

$$\frac{12c^2}{2} + 8.61 \times 0.40c - 8.61 \times 0.40 \times 4.0 = 0$$

or $c^2 + 0.574c - 2.296 = 0$, giving $c = 1.255$ in.

Effective moment of inertia

Dead load:

$$w_D = \text{self-weight of slab} = \frac{5}{12} \times 150 \text{ pcf} = 62.5 \text{ psf}$$

$$M_a = 216w = 216 \times 62.5 = 13{,}500 \text{ in.-lb} < M_{cr}$$

Hence the slab will not crack under dead load and $I_e = I_g = 125.0$ in.^{4}.

Dead load + 45% live load:

$$M_a = 216(62.5 + 0.45 \times 60) = 19{,}332 \text{ in.-lb} < M_{cr}$$

Hence the slab will not crack under dead load and 45% sustained live load and $I_e = I_g = 125.0$ in.^{4}.

Dead + live load:

$$M_a = 216(62.5 + 60.0) = 26{,}460 \text{ in.-lb} > M_{cr}$$

The section is cracked.

$$I_{cr} = \frac{bc^3}{3} + nA_s(d - c)^2 \qquad \text{from Eq. 6.7}$$

or

$$I_{cr} = \frac{12(1.255)^3}{3} + 8.61 \times 0.40(4.0 - 1.255)^2 = 33.86 \text{ in.}^4$$

$$\frac{M_{cr}}{M_a} = \frac{22{,}185}{26{,}460} = 0.838$$

$$I_e = \left(\frac{M_{cr}}{M_a}\right)^3 I_g + \left[1 - \left(\frac{M_{cr}}{M_a}\right)^3\right] I_{cr} = 0.59 \times 125.0 + 0.41 \times 33.86$$

$$= 87.63 \text{ in.}^4$$

Short-term deflection

From Table 6.3,

$$\Delta = \frac{5wl_n^2}{384E_c I_e} = \frac{5w(12.0 \times 12)^4}{384 \times 3.37 \times 10^6 I_e} \times \frac{1}{12} = \frac{0.1384}{I_e} w \text{ in.}$$

Immediate live-load deflection:

$$\Delta_L = \frac{0.1384(62.5 + 60.0)}{87.63} - \frac{0.1384(62.5)}{125.0} = 0.194 - 0.069 = 0.125 \text{ in. (3.2 mm)}$$

Immediate dead-load deflection:

$$\Delta_D = \frac{0.1384(62.5)}{125.0} = 0.069 \text{ in. } (1.8 \text{ mm})$$

Immediate 45% LL deflection:

$$\Delta_{LS} = \frac{0.1384(62.5 + 0.45 \times 60)}{125.0} - \frac{0.1384(62.5)}{125.0}$$

$$= 0.099 - 0.069 = 0.030 \text{ in. } (0.8 \text{ mm})$$

Long-term deflection

From Eq. 6.9, multiplier $\lambda = T/(1 + 50\rho')$. From Fig. 6.6, $T = 1.65$ for 24-month sustained load. Therefore

$$\lambda_\infty = 2.0 \quad \text{and} \quad \lambda_t = 1.65$$

From Eq. 6.10 the total sustained load deflection is

$$\Delta_{LT} = \Delta_L + \lambda_\infty \Delta_D + \lambda_t \Delta_{LS}$$

or

$$\Delta_{LT} = 0.125 + 2.0 \times 0.069 + 1.65 \times 0.030 = 0.313 \text{ in. } (7.9 \text{ mm})$$

Deflection requirements (Table 6.2)

$$\frac{l}{180} = \frac{12 \times 12}{180} = 0.80 \text{ in. } > \Delta_L$$

$$\frac{l}{360} = 0.40 \text{ in. } > \Delta_L$$

$$\frac{l}{480} = 0.30 \text{ in. } \simeq \Delta_{LT}$$

$$\frac{l}{240} = 0.60 \text{ in. } > \Delta_{LT}$$

Therefore the slab can support sensitive attached nonstructural elements that are otherwise damaged by large deflections.

6.7 FLEXURAL CRACKING IN BEAMS AND ONE-WAY SLABS

6.7.1 Crack-Width Evaluation

In order to evaluate the maximum crack width, the large number of variables involved, the randomness of cracking behavior, and the large degree of scatter, extensive

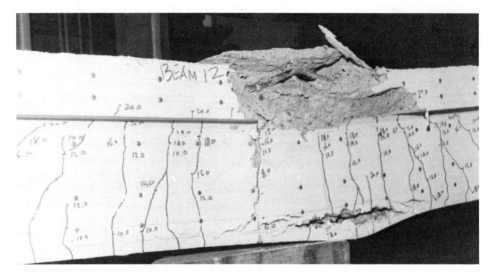

Photo 48 Flexural stabilized cracks at failure. (Tests by Nawy, et al.)

idealization and simplification are required. One simplification based on a statistical study of test data of several investigators is the Gergely–Lutz expression

$$w_{\max} = 0.076\beta f_s \sqrt[3]{d_c A} \qquad (6.11)$$

where $w_{\max}$ = crack width in units of 0.001 in. (0.0254 mm)
  $\beta = (h - c)/(d - c)$ = depth factor average value = 1.20
  d_c = thickness of cover to the center of the first layer of bars (in.)
  f_s = maximum stress (ksi) in the steel at service load level with $0.6f_y$ to be used if no computations are available
  A = area of concrete in tension divided by the number of bars (in.^{2}) = bt/γ_{bc}, where γ_{bc} is defined as the number of bars at the tension side

It is to be noted that allowance of $f_s = 0.6f_y$ in lieu of actual steel stress computations is applicable only to normal structures. Special precautions need to be taken for structures exposed to very aggressive climates, such as chemical factories or offshore structures. Additionally, the depth of the concrete area in tension in reinforced concrete is determined by having the center of gravity of the bars as the centroid of the concrete area in tension. So for a single layer of bars, the depth t of the concrete area in tension equals $2d_c$. The shaded area in Fig. 6.8 gives the total concrete area in tension.

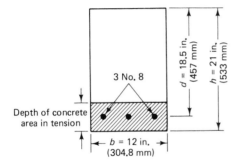

Figure 6.8   Beam geometry.

6.7.2 Example 6.5 Maximum Crack Width in a Reinforced Concrete Beam

Calculate the maximum crack width for a rectangular simply supported beam that has the cross section shown in Fig. 6.8. The beam span is 30 ft (9.14 m). It carries a working uniform load of 1000 lb/ft, including its own weight (14.6 kN/m). Given:

$$f'_c = 5000 \text{ psi, normalweight concrete (34.47 MPa)}$$

$$f_y = 60,000 \text{ psi (413.7 MPa)}$$

$$E_s = 29 \times 10^6 \text{ psi (200,000 MPa)}$$

Solution

Using $f_s = 0.6f_y$ yields

$$\beta = 1.20 \quad \text{for beams}$$

$$f_s = 0.6f_y = 0.6 \times 60.0 = 36.0 \text{ ksi}$$

$$w_{max} = 0.076 \times 1.20 \times 36.0\sqrt[3]{3.0 \times 2.4} \times 10^{-3} = 0.0137 \text{ in. } (\approx 0.35 \text{ mm})$$

6.7.3 Crack-Width Evaluation for Beams Reinforced with Bundled Bars

The bond stress between the reinforcing bars and the surrounding concrete is a major parameter affecting flexural crack spacing and hence crack width. The contact area of bundled bars is less than that of the isolated bars if they act independently. Using the perimetric reduction factor deduced from Fig. 6.9, the cracking equation becomes

$$w_{max} = 0.076\beta f_s \sqrt[3]{d'_c A'} \tag{6.12}$$

where w_{max} is the crack width in units of 0.001 in. and $A' = bt/\gamma'_{bc}$ with the factor for γ_{bc} shown in Fig. 6.9a. d'_c is the depth of cover to the center of gravity of the bundle. The steps for calculation of w_{max} are identical to those for beams reinforced with nonbundled bars.

Two bars   Three bars   Four bars
$\gamma'_{bc} = 0.815\gamma_{bc}$ $\gamma'_{bc} = 0.650\gamma_{bc}$ $\gamma'_{bc} = 0.570\gamma_{bc}$

where $A' = \dfrac{bt}{\gamma'_{bc}}$ $Z = f_s \sqrt{d'_c A'}$

(a)

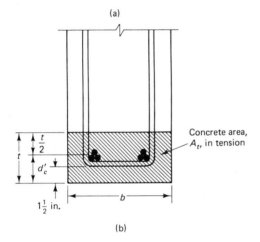

(b)

Figure 6.9 Perimetric reduction factors for beams with bundled bars: (a) perimetric factors; (b) section geometry of the concrete area in tension.

6.7.4 Example 6.6 Maximum Crack Width in a Beam Reinforced with Bundled Bars

Find the maximum flexural crack width for a reinforced concrete beam that has the cross-sectional geometry shown in Fig. 6.10. Given:

$f_y = 60,000$ psi

$f_s = 0.6 f_y = 36,000$ psi

$A_s =$ two bundles of three No. 8 bars each (25.4 mm diameter)

Size of stirrups = No. 4 (12.7 mm diameter)

Solution

$d'_c =$ center of gravity of the three bars from the outer tensions fibers

$$= (1.5 + 0.5) + \frac{2 \times 0.5 + 1 \times 1.5}{3} = 2.83 \text{ in.}$$

$t =$ depth of the concrete area in tension

$$= 2 \times 2.83 = 5.66 \text{ in.}$$

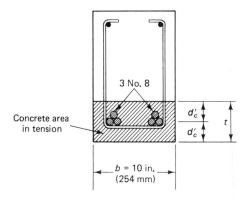

Figure 6.10 Beam geometry.

γ_{bc} = number of bars if all are of the same diameter, or the total steel area divided by the area of the largest bar if more than one size is used

= 6 in this case

$\gamma'_{bc} = 0.650\gamma_{bc} = 0.650 \times 6 = 3.9$

$A = \dfrac{bt}{\gamma'_{bc}} = \dfrac{10 \times 5.66}{3.9} = 14.51$ in.^{2}

$w_{max} = 0.076 \times 1.20 \times 36.0\sqrt[3]{2.83 \times 14.51} \times 10^{-3} = 0.011$ in. (0.29 mm)

6.8 PERMISSIBLE CRACK WIDTHS

The maximum crack width that a structural element should be permitted to develop depends on the particular function of the element and the environmental conditions to which the structure is liable to be subjected. Table 6.4 from the ACI Committee 224 report on cracking serves as a reasonable guide on permissible crack widths in concrete structures under various environmental conditions encountered. Engineering judgment is needed in determining the maximum crack width that can be tolerated.

TABLE 6.4 PERMISSIBLE CRACK WIDTHS

Exposure condition	Tolerable crack width	
	in.	mm
Dry air or protective membrane	0.016	0.41
Humidity, moist air, soil	0.012	0.30
Deicing chemicals	0.007	0.18
Seawater and seawater spray; wetting and drying	0.006	0.15
Water-retaining structures (excluding nonpressure pipes)	0.004	0.10

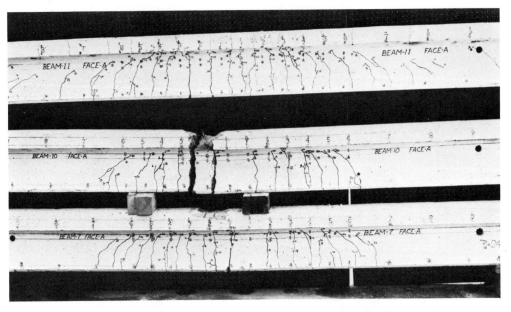

Photo 49   Typical flexural crack formation in beams. (Tests by Nawy, et al.)

6.9 CRACKING BEHAVIOR AND CRACK CONTROL IN TWO-WAY-ACTION SLABS AND PLATES

6.9.1 Crack Control Equation

The basic equation (Section 6.7) for relating crack width to strain in the reinforcement is

$$w = \alpha a_c^\beta \epsilon_s^\gamma \tag{6.13}$$

The effect of the tensile strain in the concrete between the cracks is neglected as insignificant. a_c is the crack spacing, ϵ_s the unit strain in the reinforcement, and α, β, and γ are constants. As a result of this fracture hypothesis, the mathematical model in Eq. 6.13, and the statistical analysis of the data of 90 slabs tested to failure, the following crack-control equation emerged:

$$w = K\beta f_s \sqrt{\frac{d_{b_1} s_2}{Q_{i_1}}} \tag{6.14}$$

where the quantity under the radical, $G_1 = d_{b_1} s_2 / Q_{i_1}$, is termed the grid index and can be transformed into

$$G_1 = \frac{s_1 s_2 d_c}{d_{b_1}} \frac{8}{\pi}$$

where K = fracture coefficient, having a value of $K = 2.8 \times 10^{-5}$ for uniformly loaded restrained two-way-action square slabs and plates. For concentrated loads or reactions, or when the ratio of short to long span is less than 0.75 but larger than 0.5, a value of $K = 2.1 \times 10^{-5}$ is applicable. For a span aspect ratio of 0.5, $K = 1.6 \times 10^{-5}$. Units of coefficient K are in square inch per pound.

β = ratio of the distance from the neutral axis to the tensile face of the slab to the distance from the neutral axis to the centroid of the reinforcement grid (to simplify the calculations, use $\beta = 1.25$, although it varies between 1.20 and 1.35)

f_s = actual average service load stress level, or 40% of the design yield strength, f_y (ksi)

d_{b1} = diameter of the reinforcement in direction 1 closest to the concrete outer tension fibers (in.)

s_1 = spacing of the reinforcement in direction 1.

s_2 = spacing of the reinforcement in perpendicular direction 2.

1 = direction of the reinforcement closest to the outer concrete fibers; this is the direction for which crack control check is to be made

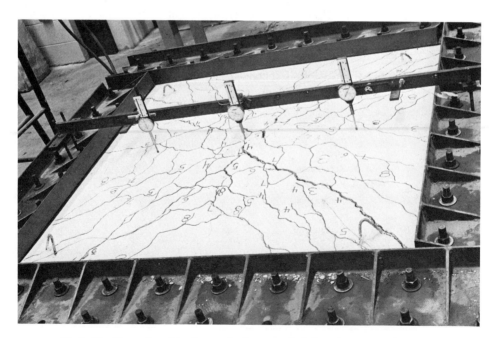

Photo 50 Flexural cracking in restrained one-panel reinforced concrete slab. (Tests by Nawy, et al.)

Q_{i_1} = active steel ratio

$$= \frac{\text{area of steel } A_s \text{ per ft width}}{12(d_{b1} + 2c_1)}$$

where c_1 is clear concrete cover measured from the tensile face of the
concrete to the nearest edge of the reinforcing bar in direction 1

w = crack width at face of concrete caused by flexural load (in.)

Subscripts 1 and 2 pertain to the directions of reinforcement. Detailed values of
the fracture coefficients for various boundary conditions are given in Table 6.5.

A graphical solution of Eq. 6.14 is given in Fig. 6.11 for

$$f_y = 60{,}000 \text{ psi } (414 \text{ MPa})$$

$$f_s = 40\% \, f_y$$

$$f_s = 24{,}000 \text{ psi } (165.5 \text{ MPa})$$

for rapid determination of the reinforcement size and spacing needed for crack control.

The grid index G_1 specifies the size and spacing of the bars in the two perpendicular directions of any concrete floor system and $w_{\max}$ is the maximum allowable
crack width.

The crack control equation and guidelines presented are important not only for
the control of corrosion in the reinforcement but also for deflection control. The
reduction of the stiffness EI of the two-way slab or plate due to orthogonal cracking

TABLE 6.5 FRACTURE COEFFICIENTS FOR SLABS AND PLATES

Loading type[a]	Slab shape	Boundary condition[b]	Span ratio,[c] S/L	Fracture coefficient, 10^{-5} K
A	Square	4 edges r	1.0	2.1
A	Square	4 edges s	1.0	2.1
B	Rectangular	4 edges r	0.5	1.6
B	Rectangular	4 edges r	0.7	2.2
B	Rectangular	3 edges r, 1 edge h	0.7	2.3
B	Rectangular	2 edges r, 2 edges h	0.7	2.7
B	Square	4 edges r	1.0	2.8
B	Square	3 edges r, 1 edge h	1.0	2.9
B	Square	2 edges r, 2 edges h	1.0	4.2

[a]Loading type: A, concentrated; B, uniformly distributed.
[b]Boundary condition: r, restrained; s, simply supported; h, hinged.
[c]Span ratio: S, clear short span; L, clear long span.

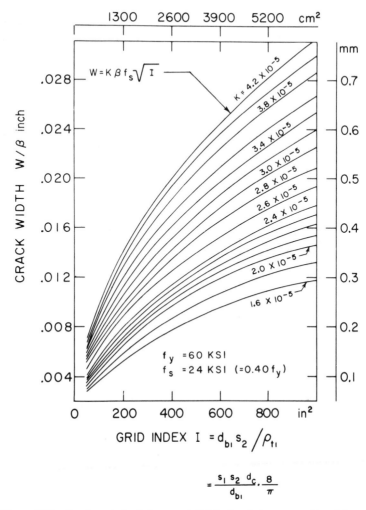

Figure 6.11 Crack control reinforcement distribution in two-way-action slabs and plates for all exposure conditions: $f_y = 60,000$ psi, $f_s = 24,000$ psi ($= 0.40 f_y$).

when the limits of permissible crack widths in Table 6.4 are exceeded can lead to excessive deflection, both short term and long term. Deflection values several times those anticipated in the design, including deflection due to construction loading, can be reasonably controlled through camber and control of the flexural crack width in the slab or plate. Proper selection of the *reinforcement spacing* s_1 and s_2 in both perpendicular directions as discussed in this section, and not exceeding 12 in. center to center, can maintain good serviceability performance of a slab system under normal and reasonable overload conditions. The 1985 Australian Code and other codes generally follow the principles presented on the selection of reinforcement size and spacing in slabs and plates and good engineering practice mandates such caution.

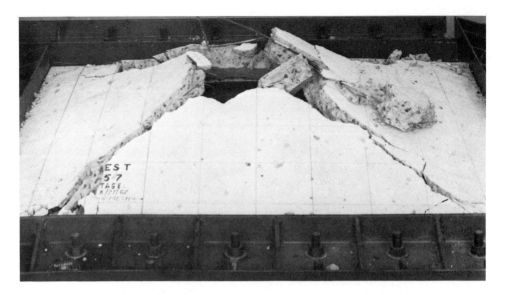

Photo 51 Rectangular concrete slab at rupture. (Tests by Nawy, et al.)

6.9.2 Example 6.7: Crack Control Evaluation for Serviceability in a Rectangular Panel Subjected to Severe Exposure Conditions

Select the bar size and spacing necessary for crack control at the column reaction region of a 7-in.-thick slab shown in Fig. 6.12 that is uniformly loaded. Select the bar size for two conditions:

Condition A: Floor is subjected to severe exposure of humidity and moist air.

Condition B: Floor sustains an aggressive chemical environment where the design working stress level in the reinforcement is limited to 15 ksi (15,000 psi).

Given:

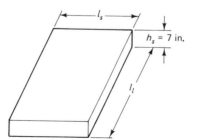

Figure 6.12 Rectangular panel.

$$\beta = 1.20$$

$$l_s/l_l = 0.8$$

$$f_y = 60 \text{ ksi (414 Mpa)}$$

Solution

Condition A: Humidity and moist air

Permissible $w_{max} = 0.012$ in. (0.3 mm) (Table 6.4). Try No. 4 bars $d_b = 0.5$, $d_c = 0.75 + 0.25 = 1.0$ in. Assume that $s_1 = s_2 = s$ for the given panel. The aspect ratio $l_s/l_l = 0.8$. $K = 2.1 \times 10^{-5}$ for concentrated reaction at the column support (Table 6.5).

$$0.012 = 2.1 \times 10^{-5} \times 1.20 \times 0.4 \times 60\sqrt{G_I}$$

to give $G_I = 394$ in.^{2}. Therefore

$$394 = \frac{s^2 d_c}{d_{b1}} \times \frac{8}{\pi} = \frac{s^2 \times 1.0}{0.5} \times \frac{8}{\pi}$$

$$s = 8.8 \text{ in.}$$

So use No. 4 bars at $8\frac{1}{2}$ in. center to center each way for crack control.

Condition B: Aggressive chemical environment

Permissible $w_{max} = 0.007$ in. (0.18 mm) (Table 6.4) $f_s = 15$ ksi to be used as a low stress level for sanitary or water-retaining structures instead of $0.4f_y$. Try No. 5 bars ($d_{b1} = 0.625$ in.)

$$0.007 = 2.1 \times 10^{-5} \times 1.20 \times 15.0\sqrt{G_I}$$

to give a grid index $G_I = 343$ in.^{2}

$$d_{c1} = 0.75 + 0.312 = 1.06 \text{ in.}$$

$$G_I = 343 = \frac{s^2 \times 1.06}{0.625} \times \frac{8}{\pi} \qquad \text{to get } s = 8.9 \text{ in.}$$

Use No. 5 bars at 9 in. (229 mm) center-to-center spacing each way for crack control.

Reinforcement Summary

Condition A: No. 4 bars at $8\frac{1}{2}$ in. c-c (12.7 mm diameter at 216 mm c-c)
Condition B: No. 5 bars at 9 in. c-c (15.9 mm diameter at 229 mm c-c)

SELECTED REFERENCES

6.1   Branson, D. E., *Deformation of Concrete Structures,* McGraw-Hill, New York, 1977.

6.2   Gergely, P., and L. A. Lutz, *Maximum Crack Width in Reinforced Concrete Flexural Members,* Special Publication SP20, American Concrete Institute, Detroit, 1968.

6.3   Nawy, E. G., "Crack Control in Beams Reinforced with Bundled Bars," *Journal of the American Concrete Institute,* Proc. Vol. 69, October 1972, pp. 637–640.

6.4 Nawy, E. G., "Crack Control in Reinforced Concrete Structures," *Journal of the American Concrete Institute,* Proc. Vol. 65, October 1968, pp. 825–838.

6.5 Nawy, E. G., "Crack Control through Reinforcement Distribution in Two-Way Acting Slabs and Plates," *Journal of the American Concrete Institute,* Proc. Vol. 69, No. 4, April 1972, pp. 217–219.

6.6 Nawy, E. G. and K. W. Blair, "Further Studies on Flexural Crack Control in Structural Slab Systems;" "Discussion" by ACI Code Committee 318; and "Authors' Closure," *Journal of the American Concrete Institute,* Proc. Vol. 70, January 1973, pp. 61–63.

6.7 Nawy, E. G., "Reinforced Concrete—A Fundamental Approach," Prentice-Hall, Inc., Englewood Cliffs, NJ, 1985, 720 pp.

6.8 Nawy, E. G., "Strength, Serviceability and Ductility," Chapter 12 in *Handbook of Structural Concrete,* Pitman Books, London/McGraw-Hill, New York, 1983, 1968 pp.

6.9 ACI Committee 224, "Control of Cracking in Concrete Structures," *Journal of the American Concrete Institute,* Proc. Vol. 77, October 1980, pp. 35–76; Proc. Vol. 69, December 1972, pp. 717–753.

6.10 ACI Committee 340, *Design Hand Book: Beams, Slabs, Etc.,* Special Publication SP17 (84), American Concrete Institute, Detroit, Vol. 1, 1984, 374 pp.

6.11 ACI Committee 435, "Variability of Deflections of Simply Supported Reinforced Concrete Beams," *Journal of the American Concrete Institute,* Proc. Vol. 69, January 1972, pp. 29–35.

PROBLEMS FOR SOLUTION

6.1 Calculate I_g and I_{cr} for cross sections (a) through (f) in Fig. 6.13. Given:

$f_c = 4000$ psi (27.58 Mpa), normalweight concrete

$f_y = 60,000$ psi (413.7 MPa)

$E_s = 29 \times 10^6$ psi (200,000 MPa)

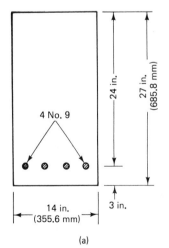

(a)

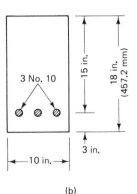

(b)

Figure 6.13 Beam cross sections for deflection calculations.

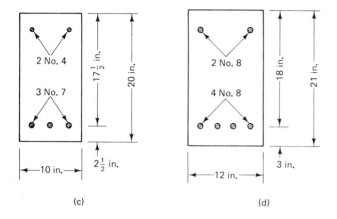

(c)   (d)

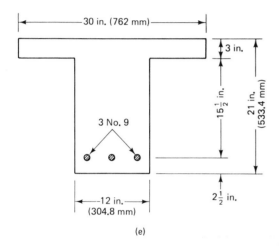

(e)

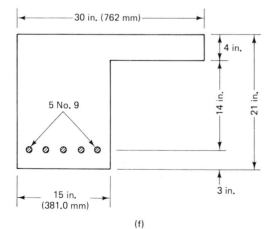

(f)   Figure 6.13 (cont.)

6.2   Calculate the deflection due to dead load, and dead load plus live load, for the following cases in Problem 6.1: cross sections (a), (d), and (e). Use for service load levels $0.2M_u$ as maximum dead load moment and $0.35\,M_u$ as maximum live load moment. Assume that all beams are simply supported and have a span of 22 ft (6.71 m).

6.3   Calculate the maximum immediate and long-term deflection for a 6-in.-thick slab on simple supports spanning over 13 ft. The service dead and live loads are 70 psf (33.52 kPa) and 120 psf (57.46 kPa), respectively. The reinforcement consists of No. 5 bars (16 mm diameter) at 6 in. center to center (154.2 mm center to center). Also, check which limitations, if any, need to be placed on its usage. Assume that 60% of the live load is sustained over a 30-month period. Given:

$$f_c' = 4500 \text{ psi (31.03 MPa)}$$

$$E_s = 29 \times 10^6 \text{ psi (200,000 MPa)}$$

6.4   A rectangular beam under simple bending has the dimensions shown in Fig. 6.14. It is subjected to an aggressive chemical environment. Calculate the maximum expected flexural crack width and whether the beam satisfies the serviceability criteria for crack control. Given:

$$f_c' = 5000 \text{ psi (34.47 MPa)}$$

$$f_y = 60,000 \text{ psi (413.7 MPa)}$$

Minimum clear cover $= 1\tfrac{1}{2}$ in. (38.1 mm)

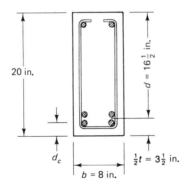

Figure 6.14   Beam geometry.

6.5   Calculate the maximum crack width in a two-way interior panel of a reinforced concrete floor system. The slab thickness is 8 in. (203.2 mm) and the panel size is 20 ft × 28 ft (6.10 m × 8.53 m). Also, design the size and spacing of the reinforcement necessary for crack control, assuming that

(a) The floor is exposed to normal environment.

(b) The floor is part of a parking garage.

Given

$$f_y = 60 \text{ ksi (414 MPa)}$$

CHAPTER

7

Columns

7.1 INTRODUCTION

Columns are vertical compression members of a structural frame intended to support the load-carrying beams. They transmit loads from the upper floors to the lower levels and then to the soil through the foundations. Because columns are compression elements, failure of one column in a critical location can cause the progressive collapse of the adjoining floors and the ultimate total collapse of the entire structure.

Structural column failure is of major significance in terms of economic as well as human loss. Thus extreme care needs to be taken in column design, with a higher reserve strength than in the case of beams and other horizontal structural elements, particularly because compression failure provides little visual warning. As will be seen in subsequent sections, the ACI code requires a considerably lower strength reduction factor ϕ in the design of compression members than the ϕ factors in flexure, shear, or torsion.

The principles of stress and strain compatibility used in the analysis (design) of beams discussed in Chapter 3 are equally applicable to columns. A new factor is introduced, however: the addition of an external axial force to the bending moments acting on the critical section. Consequently, an adjustment to the force and moment equilibrium equations developed for beams becomes necessary to account for combined compression and bending.

The amount of reinforcement in the case of beams was controlled so as to have ductile failure behavior. In the case of columns, the axial load will occasionally dominate; thus compression failure behavior in cases of a large axial load/bending moment ratio cannot be avoided.

195

As the load on a column continues to increase, cracking becomes more intense along the height of the column at the transverse tie location. At the limit state of failure, the concrete cover in tied columns or the shell of concrete outside the spirals of spirally confined columns spalls and the longitudinal bars become exposed. Additional load leads to failure and local buckling of the individual longitudinal bars at the unsupported length between the ties. It is noted that, at the limit state of failure, the concrete cover to the reinforcement spalls first after the bond is destroyed.

As in the case of beams, the strength of columns is evaluated on the basis of the following principles.

1. A linear strain distribution exists across the thickness of the column.
2. There is no slippage between the concrete and the steel (i.e., the strain in steel and in the adjoining concrete is the same).
3. The maximum allowable concrete strain at failure for the purpose of strength calculations = 0.003 in./in.
4. The tensile resistance of the concrete is negligible and is disregarded in computations.

7.2 TYPES OF COLUMNS

Columns can be classified on the basis of the form and arrangement of reinforcement, the position of the load on the cross section, and the length of the column in relation to its lateral dimensions.

The form and arrangement of the reinforcement identify three types of columns, as shown in Fig. 7.1:

1. Rectangular or square columns reinforced with longitudinal bars and lateral ties (Fig. 7.1a).
2. Circular columns reinforced with longitudinal reinforcement and spiral reinforcement, or lateral ties (Fig. 7.1b).
3. Composite columns in which steel structural shapes are encased in concrete. The structural shapes could be placed inside the reinforcement cage, as shown in Fig. 7.1c.

Although tied columns are the most commonly used because of lower construction costs, spirally bound rectangular or circular columns are also used where increased ductility is needed, such as in earthquake zones. The ability of the spiral column to sustain the maximum load at excessive deformations prevents the complete collapse of the structure before the total redistribution of moments and stresses is complete. Figure 7.2 shows the large increase in ductility (toughness) due to the effect of spiral binding.

Based on the position of the load on the cross section, columns can be classified as concentrically or eccentrically loaded, as shown in Fig. 7.3. Concentrically loaded

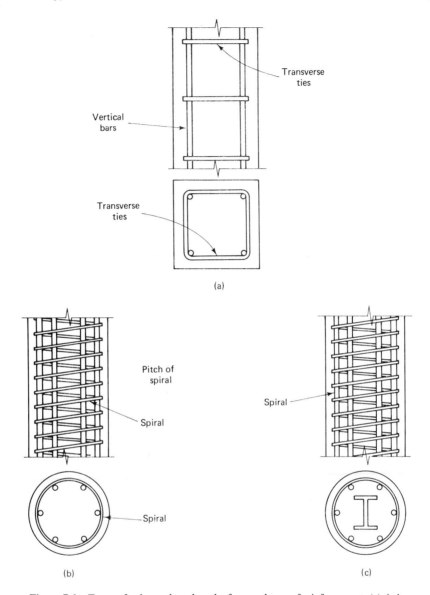

Figure 7.1   Types of columns based on the form and type of reinforcement: (a) tied column; (b) spiral column; (c) composite column.

columns (Fig. 7.3a) carry no moment. In practice, however, all columns need to be designed for some unforeseen or accidental eccentricity due to such causes as imperfections in the vertical alignment of formwork.

Eccentrically loaded columns (Fig. 7.3b and c) are subjected to moment in addition to the axial force. The moment can be converted to a load P and an eccen-

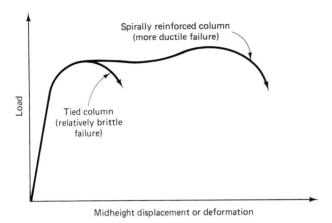

Figure 7.2 Comparison of load-deformation behavior of tied and spirally bound columns.

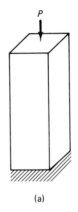

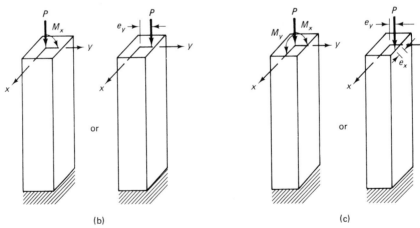

Figure 7.3 Types of columns based on the position of the load on the cross section: (a) concentrically loaded column; (b) axial load plus uniaxial moment; (c) axial load plus biaxial moment.

tricity e, as shown in Fig. 7.3b and c. The moment can be uniaxial, as in the case of an exterior column in a multistory building frame, or when two adjacent panels are not similarly loaded, such as columns A and B in Fig. 7.4. A column is considered biaxially loaded when bending occurs about both the x and y axes, such as in the case of corner column C of Fig. 7.4.

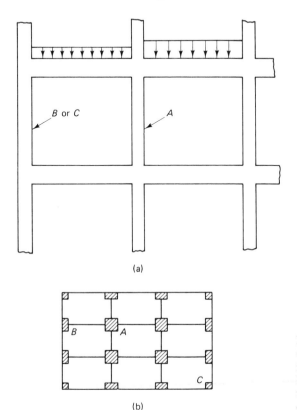

(a)

(b)

Figure 7.4 Bending of columns: (a) frame elevation; (b) framing plan, A, interior column under nonsymmetrical load-uniaxial bending; B, exterior column uniaxial bending; C, exterior corner column, biaxial bending.

Failure of columns could occur as a result of material failure by initial yielding of the steel at the tension face or initial crushing of the concrete at the compression face, or by loss of lateral structural stability (i.e., through buckling).

If a column fails due to initial material failure, it is classified as a *short column*. As the length of the column increases, the probability that failure will occur by buckling also increases. Therefore the transition from the short column (material failure) to the long column (failure due to buckling) is defined by using the ratio of the effective length kl_u to the radius of gyration r. The height l_u, is the unsupported length of the column and k is a factor that depends on end conditions of the column and whether it is braced or unbraced. In the case of unbraced columns, for example, if kl_u/r is less than or equal to 22, such a column is classified as a short column, in accordance with the ACI load criteria. Otherwise it is defined as a long or a slender column. The ratio kl_u/r is called the *slenderness ratio*.

7.3 STRENGTH OF SHORT CONCENTRICALLY LOADED COLUMNS

Consider a column of gross cross-sectional area A_g with width b and total depth h, reinforced with a total area of steel A_{st} on all faces of the column. The net cross-sectional area of the concrete is $A_g - A_{st}$.

Figure 7.5 presents the stress history in the concrete and the steel as the column load is increased. Both the steel and the concrete behave elastically at first. At a strain of approximately 0.002 in./in. to 0.003 in./in., the concrete reaches its maximum strength f'_c. Theoretically the maximum load that the column can take occurs when the stress in the concrete reaches f'_c. Further increase is possible if strain hardening occurs in the steel at about 0.003 in./in. strain levels.

Therefore the maximum concentric load capacity of the column can be obtained by adding the contribution of the concrete, which is $(A_g - A_{st})0.85f'_c$, and the contribution of the steel, which is $A_{st}f_y$, where A_g is the total gross area of the concrete section and A_{st} is the total steel area $= A_s + A'_s$. The value of $0.85f'_c$ instead of f'_c is

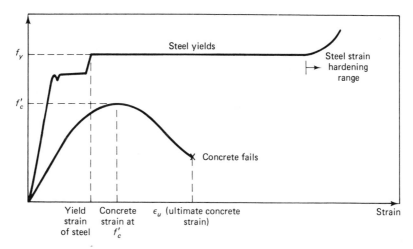

Figure 7.5 Stress-strain behavior of concrete and steel (concentric load).

used in the calculation because it is found that the maximum attainable strength in the actual structure approximates $0.85f_c'$. Thus the nominal concentric load capacity, P_0, can be expressed as

$$P_0 = 0.85f_c'(A_g - A_{st}) + A_{st}f_y \tag{7.1}$$

It should be noted that the concentric load causes uniform compression throughout the cross section. Consequently, at failure, the strain and stress will be uniform across the cross section, as shown in Fig. 7.6.

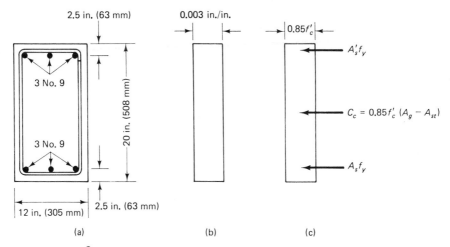

Figure 7.6 Column geometry—strain and stress diagrams (concentric load): (a) cross section; (b) concrete strain; (c) stress (forces).

Attaining zero eccentricity in actual structures is highly improbable. Eccentricities could easily develop because of such factors as slight inaccuracies in the layout of columns and unsymmetric loading due to the difference in thickness of the slabs in adjacent spans or imperfections in the alignment, as indicated earlier. So a minimum eccentricity of 10% of the thickness of the column in the direction perpendicular to its axis of bending is considered an acceptable assumption for columns with ties and 5% for spirally reinforced columns.

To reduce the calculations necessary for analysis and design for minimum eccentricity, the ACI code specifies a reduction of 20% in the axial load for tied columns and a 15% reduction for spiral columns. Using these factors, the maximum nominal axial load capacity of columns cannot be taken greater than

$$P_{n(\max)} = 0.8[0.85f'_c(A_g - A_{st}) + A_{st}f_y] \qquad (7.2)$$

for tied reinforced columns and

$$P_{n(\max)} = 0.85[0.85f'_c(A_g - A_{st}) + A_{st}f_y] \qquad (7.3)$$

for spirally reinforced columns.

These nominal loads should be reduced further by using strength reduction factors ϕ, as explained in later sections. Normally, for design purposes, $A_g - A_{st}$ can be assumed to be equal to A_g without great loss in accuracy.

7.3.1 Example 7.1: Analysis of an Axially Loaded Short Rectangular Tied Column

A short tied column is subjected to axial load only. It has the geometry shown in Fig. 7.6a and is reinforced with three No. 9 bars (28.6 mm diameter) on each of the two faces parallel to the x axis of bending. Calculate the maximum nominal axial load strength $P_{n(\max)}$. Given:

$$f'_c = 4000 \text{ psi (27.6 MPa)}$$

$$f_y = 60,000 \text{ psi (414 MPa)}$$

Solution

$$A_s = A'_s = 3 \text{ in.}^2$$

Therefore $A_{st} = 6$ in.^{2} Using Eq. 7.2 yields

$$P_{n(\max)} = 0.8\{0.85 \times 4000[(12 \times 20) - 6] + 6 \times 60,000\}$$

$$= 924,480 \text{ lb (4110 kN)}$$

If $A_g - A_{st}$ is taken as equal to A_g, it results in

$$P_{n(\max)} = 0.8(0.85 \times 4000 \times 12 \times 20 + 6 \times 60,000)$$

$$= 940,800 \text{ lb (4180 kN)}$$

Note from Fig. 7.6b and c that the entire concrete cross section is subjected to a uniform compressive stress of $0.85f'_c$ and a uniform strain of 0.003 in./in.

7.3.2 Example 7.2: Analysis of an Axially Loaded Short Circular Column

A 20-in.-diameter short spirally reinforced circular column is symmetrically reinforced with six No. 8 bars, as shown in Fig. 7.7. Calculate the strength $P_{n(\max)}$ of this column if subjected to axial load only. Given:

$$f'_c = 4000 \text{ psi } (27.6 \text{ MPa})$$

$$f_y = 60,000 \text{ psi } (414 \text{ MPa})$$

Solution

$$A_{st} = 4.74 \text{ in.}^2$$

$$A_g = \frac{\pi}{4}(20)^2 = 314 \text{ in.}^2$$

Using Eq. 7.3 yields

$$P_{n(\max)} = 0.85[0.85 \times 4000(314 - 4.74) + 4.74 \times 60,000]$$

$$= 1,135,501 \text{ lb } (5065 \text{ kN})$$

or assuming that $A_g - A_{st} \cong A_g$,

$$P_{n(\max)} = 0.85(0.85 \times 4000 \times 314 + 4.74 \times 60,000)$$

$$= 1,149,200 \text{ lb } (5062 \text{ kN})$$

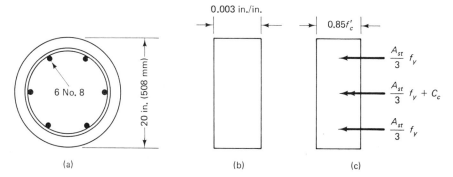

Figure 7.7 Column geometry—strain and stress diagrams (concentric load): (a) cross section; (b) concrete strain; (c) stress (forces).

7.4 STRENGTH OF ECCENTRICALLY LOADED COLUMNS: AXIAL LOAD AND BENDING

7.4.1 Behavior of Eccentrically Loaded Short Columns

The same principles concerning the stress distribution and the equivalent rectangular stress block applied to beams are equally applicable to columns. Figure 7.8 shows a typical rectangular column cross section with strain, stress, and force distribution diagrams. The diagram differs from Fig. 3.13 in the introduction of an additional longitudinal nominal force P_n at the limit failure state acting at an eccentricity e from the plastic (geometric) centroid of the section. The depth of the neutral axis primarily determines the strength of the column.

The equilibrium expressions for forces and moments from Fig. 7.8 can be expressed as follows for nonslender columns.

$$\text{Nominal axial resisting force } P_n \text{ at failure} = C_c + C_s - T_s \qquad (7.4)$$

Nominal resisting moment M_n, which is equal to $P_n e$, can be obtained by writing the moment equilibrium equation about the plastic centroid. For columns with symmetrical reinforcement, the plastic centroid is the same as the geometric centroid.

$$M_n = P_n e = C_c\left(\bar{y} - \frac{a}{2}\right) + C_s(\bar{y} - d') + T_s(d - \bar{y}) \qquad (7.5)$$

since

$$C_c = 0.85 f'_c ba$$

$$C_s = A'_s f'_s$$

$$T_s = A_s f_s$$

Equations 7.4 and 7.5 can be rewritten

$$P_n = 0.85 f'_c ba + A'_s f'_s - A_s f_s \qquad (7.6)$$

$$M_n = P_n e = 0.85 f'_c ba\left(\bar{y} - \frac{a}{2}\right) + A'_s f'_s(\bar{y} - d') - A_s f_s(d - \bar{y}) \qquad (7.7)$$

In Eqs. 7.6 and 7.7 the depth of the neutral axis c is assumed to be less than the effective depth d of the section and the steel at the tension face is in actual tension. Such a condition changes if the eccentricity e of the axial force P_n is very small. For such small eccentricities where the total cross section is in compression, contribution of the tension steel should be added to the contribution of concrete and compression steel. The term $A_s f_s$ in Eqs. 7.6 and 7.7 in such a case would have a positive sign, for all the steel is in compression. It is also assumed that $(ba - A'_s) \simeq ba$; that is, the volume of concrete displaced by compression steel is negligible.

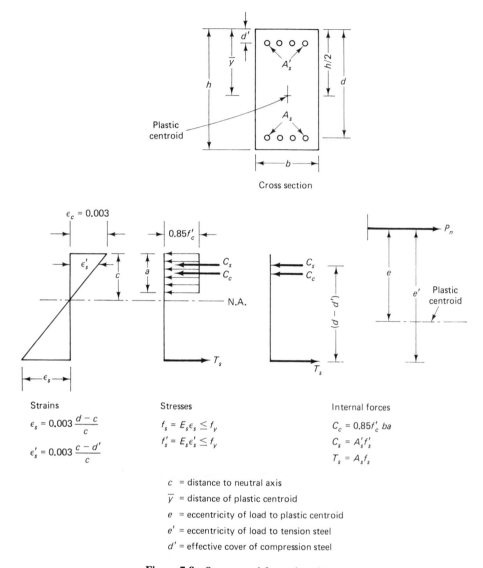

$$\epsilon_c = 0.003$$

Strains

$$\epsilon_s = 0.003 \frac{d-c}{c}$$

$$\epsilon_s' = 0.003 \frac{c-d'}{c}$$

Stresses

$$f_s = E_s \epsilon_s \leq f_y$$

$$f_s' = E_s \epsilon_s' \leq f_y$$

Internal forces

$$C_c = 0.85 f_c' \, ba$$

$$C_s = A_s' f_s'$$

$$T_s = A_s f_s$$

c = distance to neutral axis

$\bar{y}$ = distance of plastic centroid

e = eccentricity of load to plastic centroid

e' = eccentricity of load to tension steel

d' = effective cover of compression steel

Figure 7.8   Stresses and forces in columns.

If a computer is used for the analysis (design), more refined solutions can be obtained with minimum effort. It can be observed that the error in neglecting the contribution of the displaced concrete is not significant.

It should be noted that the axial force P_n cannot exceed the maximum axial load strength $P_{n(\text{max})}$, calculated by using Eq. 7.2. Depending on the magnitude of the eccentricity e, the compression steel A_s' or the tension steel A_s will reach the yield strength, f_y. Stress f'_s reaches f_y when failure occurs by crushing of the concrete. If failure develops by yielding of the tension steel, f_s should be replaced by f_y. When the magnitude of f'_c or f_s is less than f_y, the actual stresses can be calculated by using the following equations, which are obtained from similar triangles in the strain distribution across the depth of the section (Fig. 7.8).

$$f'_s = E_s\epsilon'_s = E_s\frac{0.003(c - d')}{c} \leq f_y \tag{7.8}$$

$$f_s = E_s\epsilon_s = E_s\frac{0.003(d - c)}{c} \leq f_y \tag{7.9}$$

7.4.2 Basic Column Equations 7.6 and 7.7 and Trial-and-Adjustment Procedure for Analysis (Design) of Columns

Equations 7.6 and 7.7 determine the nominal axial load P_n that can be safely applied at an eccentricity e for any eccentrically loaded column. If one examines these two expressions, the following unknowns can be identified:

1. Depth of the equivalent stress block, a
2. Stress in compression steel, f'_s
3. Stress in tension steel, f_s
4. P_n for the given e or vice versa

The stresses f'_s and f_s can be expressed in terms of the depth of neutral axis c as in Eqs. 7.8 and 7.9 and thus in terms of a. The two remaining unknowns, a and P_n, can be solved by using Eqs. 7.6 and 7.7. Combining Eqs. 7.6 to 7.9, however, leads to a cubical equation in terms of the neutral axis depth c. One also has to check whether the steel stresses are less than the yield strength f_y. Thus the following trial-and-adjustment procedure is suggested for a general case of analysis (design).

Assume a value for the distance c down to the neutral axis for a given section geometry and eccentricity e. This value is a measure of the compression block depth a because $a = \beta_1 c$. Taking the assumed value of c, calculate the axial load P_n by using Eq. 7.6 and $a = \beta_1 c$. Calculate the stresses f'_s and f_s in compression and tension steel, respectively, using Eqs. 7.8 and 7.9. Also, calculate the eccentricity corresponding to the calculated load P_n, using Eq. 7.7. This calculated eccentricity should match the given eccentricity e. If not, repeat the steps until a convergence is accomplished. If the calculated eccentricity is larger than the given eccentricity, such a result indicates that the assumed value for c and the corresponding depth a of the compression block are less than the actual depth. In such a case, try another cycle, assuming a larger value of c.

This trial-and-adjustment process converges rapidly and becomes exceedingly simpler if a computer program is used. This discussion pertains to a general case. Simplifying assumptions can be made in most cases to shorten the iteration process.

7.5 MODES OF MATERIAL FAILURE IN COLUMNS

Based on the magnitude of strain in the steel reinforcement at the tension side (Fig. 7.8), the section is subjected to one of two initial conditions of failure as follows:

1. Tension failure by initial yielding of steel at the tension side
2. Compression failure by initial crushing of the concrete at the compression side

The balanced condition occurs when failure develops simultaneously in tension and compression.

If P_n is the axial load and P_{nb} is the axial load corresponding to the balanced condition, then

$$P_n < P_{nb} \quad \text{tension failure}$$

$$P_n = P_{nb} \quad \text{balanced failure}$$

$$P_n > P_{nb} \quad \text{compression failure}$$

In all these cases, the strain-compatibility relationship must be maintained.

7.5.1 Balanced Failure in Rectangular Column Sections

As the eccentricity decreases, a gradual transition from a primary tension failure to a primary compression failure takes place. The balanced failure condition is reached when the tension steel reaches its yield strain ϵ_y at precisely the same load level as the concrete reaches its ultimate strain ϵ_c (0.003 in./in.) and starts crushing.

Using similar triangles, an expression for the depth of neutral axis c_b at balanced condition can be written (Fig. 7.8)

$$\frac{c_b}{d} = \frac{0.003}{0.003 + f_y/E_s} \tag{7.10a}$$

or using $E_s = 29 \times 10^6$ psi,

$$c_b = d\frac{87,000}{87,000 + f_y} \tag{7.10b}$$

$$a_b = \beta_1 c_b = \beta_1 d\frac{87,000}{87,000 + f_y} \tag{7.11}$$

The axial load corresponding to balanced condition P_{nb} and the corresponding eccentricity e_b can be determined by using this a_b in Eqs. 7.6 and 7.7.

$$P_{nb} = 0.85f'_c ba_b + A'_s f'_s - A_s f_y \tag{7.12}$$

$$M_{nb} = P_{nb}e_b = 0.85f'_c ba_b\left(\bar{y} - \frac{a_b}{2}\right) + A'_s f'_s(\bar{y} - d') + A_s f_y(d - \bar{y}) \tag{7.13}$$

where
$$f'_s = 0.003E_s\frac{c_b - d'}{c_b} \le f_y \tag{7.14}$$

and $\bar{y}$ is the distance from the compression fibers to the plastic or geometric centroid. Note that because a_b and f'_s are known, both P_{nb} and e_b can be calculated without going through any trial runs. If $A'_s = A_s$, then $\bar{y} = 0.5h$.

7.5.2 Example 7.3: Analysis of a Column Subjected to Balanced Failure

Calculate the nominal balanced load, P_{nb}, in Ex. 7.1 and the corresponding eccentricity, e_b, for the balanced failure condition if the column shown in Fig. 7.9 is subjected to combined bending and axial load. Given:

$b = 12$ in.

$d = 17.5$ in.

$h = 20$ in.

$d' = 2.5$ in.

$A_s = A'_s = 3.0$ in.^{2}

$f'_c = 4000$ psi

$f_y = 60,000$ psi

Solution

Using Eq. 7.10b gives

$$c_b = 17.5\left(\frac{87,000}{87,000 + 60,000}\right) = 10.36 \text{ in.}$$

$$a_b = \beta_1 c_b = 0.85 \times 10.36 = 8.81 \text{ in.}$$

$$f'_s = 0.003E_s\frac{c_b - d'}{c_b} \le f_y$$

$$= (0.003)(29 \times 10^6)\left(\frac{10.36 - 2.5}{10.36}\right) = 66,006 \text{ psi} > f_y$$

Therefore

$$f'_s = f_y = 60,000 \text{ psi}$$

Using Eq. 7.12, we have

$$P_{nb} = 0.85 \times 4000 \times 12 \times 8.81 + 3 \times 60,000 - 3 \times 60,000$$

$$= 359,448 \text{ lb}$$

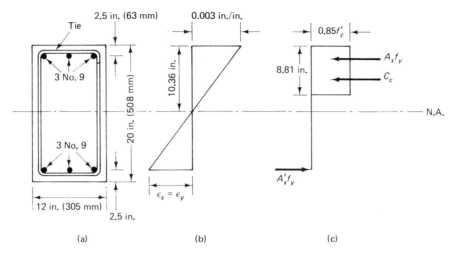

Figure 7.9 Column geometry—strain and stress diagrams (balanced failure):
(a) cross section; (b) balanced strains; (c) stress.

Using Eq. 7.13 and $\bar{y} = h/2 = 10$ in. yields

$$M_{nb} = 0.85 \times 4000 \times 12 \times 8.81\left(10 - \frac{8.81}{2}\right) + 3 \times 60,000(10 - 2.5) + 3$$

$$\times 60,000(17.5 - 10)$$

$$= 4,711,111 \text{ in.-lb } (532 \text{ kN-m})$$

$$e_b = \frac{M_{nb}}{P_{nb}} = \frac{4,711,111}{359,448} = 13.1 \text{ in. } (333 \text{ mm})$$

(If displaced concrete is taken into account in the calculation, $P_{nb} = 348,986$ lb and
$e_b = 13.3$ in.)

7.5.3 Tension Failure in Rectangular Column Sections

The initial limit state of failure in cases of large eccentricity occurs by yielding of steel
at the tension side. The transition from compression failure to tension failure takes
place at $e = e_b$. If e is larger than e_b or $P_n < P_{nb}$, the failure will be in tension through
initial yielding of the tensile reinforcement. Equations 7.6 and 7.7 are applicable in
the analysis (design) by substituting the yield strength f_y for the stress f_s in the tension
reinforcement. The stress f_s' in the compression reinforcement may or may not be the
yield strength and the actual stress f_s' should be calculated by using Eq. 7.8.

Symmetrical reinforcement is usually used such that $A_s' = A_s$ in order to prevent
the possible interchange of the compression reinforcement with the tension reinforce-

Photo 53 Eccentrically loaded column at limit state of failure. (Tests by Nawy, et al.)

ment during bar cage placement. Symmetry of reinforcement is also often necessary when the possibility of stress reversal due to change in wind direction exists.

If the compression steel is assumed to have yielded, and $A_s = A_s'$, Eqs. 7.6 and 7.7 can be rewritten

$$P_n = 0.85f_c'ba \qquad (7.15)$$

$$M_n = P_ne = 0.85f_c'ba\left(\bar{y} - \frac{a}{2}\right) + A_s'f_y(\bar{y} - d') + A_sf_y(d - \bar{y}) \quad (7.16a)$$

or $$M_n = P_ne = 0.85f_c'ba\left(\frac{h}{2} - \frac{a}{2}\right) + A_sf_y(d - d') \qquad (7.16b)$$

In Eq. 7.16b the plastic (geometric) centroid is replaced by $h/2$ for symmetrical reinforcement and A_s' is replaced by A_s.

Additionally, Eqs. 7.15 and 7.16b can be combined to obtain a single equation for P_n. Replacing $0.85f_c'ba$ in Eq. 7.16b by Eq. 7.15 gives

$$P_ne = P_n\left(\frac{h}{2} - \frac{a}{2}\right) + A_sf_y(d - d') \qquad (7.16c)$$

Because $a = P_n/0.85f_c'b$ from Eq. 7.15,

$$P_n e = P_n\left(\frac{h}{2} - \frac{P_n}{1.7f_c'b}\right) + A_s f_y(d - d') \tag{7.16d}$$

$$\frac{P_n^2}{1.7f_c'b} - P_n\left(\frac{h}{2} - e\right) - A_s f_y(d - d') = 0 \tag{7.16e}$$

If

$$\rho = \rho' = \frac{A_s}{bd} \tag{7.16f}$$

$$P_n = 0.85f_c'b\left[\left(\frac{h}{2} - e\right) + \sqrt{\left(\frac{h}{2} - e\right)^2 + \frac{2A_s f_y(d - d')}{0.85f_c'b}}\right] \tag{7.17}$$

and if

$$m = \frac{f_y}{0.85f_c'} \tag{7.18}$$

then Eq. 7.16e can be rewritten

$$P_n = 0.85f_c'bd\left[\frac{h - 2e}{2d} + \sqrt{\left(\frac{h - 2e}{2d}\right)^2 + 2m\rho\left(1 - \frac{d'}{d}\right)}\right] \tag{7.19}$$

Replacing the eccentricity e (distance between the plastic centroid and the load) with e' (distance between the tension steel and the load), Eq. 7.19 can also be rewritten

$$P_n = 0.85f_c'bd\left[\left(1 - \frac{e'}{d}\right) + \sqrt{\left(1 - \frac{e'}{d}\right)^2 + 2m\rho\left(1 - \frac{d'}{d}\right)}\right] \tag{7.20}$$

Note that $e' = [e + (d - h/2)]$ in Fig. 7.8 and

$$\frac{h - 2e}{2d} = 1 - \frac{e'}{d}$$

In nonstandard cases where the reinforcement is not symmetrical (i.e., if ρ is not equal to ρ') and if the concrete displaced by compression steel is considered, the compressive force contribution of concrete, C_c, is changed from $0.85f_c'ba$ to $0.85f_c'(ba - A_s')$ in Eqs. 7.15 and 7.16a, Eq. 7.19 then changes to

$$P_n = 0.85f_c'bd\left[\rho'(m - 1) - \rho m + \left(1 - \frac{e'}{d}\right)\right]$$

$$+ \sqrt{\left(1 - \frac{e'}{d}\right)^2 + 2\left[\frac{e'}{d}(\rho m - \rho'm + \rho') + \rho'(m - 1)\left(1 - \frac{d'}{d}\right)\right]} \tag{7.21}$$

where e' is the distance between the axial force P_n and the tension steel (or eccentricity to tension steel).

$$\rho = \frac{A_s}{bd} \tag{7.22a}$$

$$\rho' = \frac{A'_s}{bd} \tag{7.22b}$$

Equations 7.20 and 7.21 are valid only if the compression steel yields. Otherwise Eqs. 7.6, 7.7, and 7.8 should be used for obtaining P_n. Examples 7.4 and 7.5 illustrate the design process of a column controlled by initial tension failure.

7.5.4 Example 7.4: Analysis of a Column Controlled by Tension Failure; Stress in Compression Steel Equals Yield Strength

Calculate the nominal axial load strength P_n of the section in Ex. 7.1 (see Fig. 7.10) if the load acts at an eccentricity $e = 14$ in. (356 mm). Given:

$b = 12$ in.

$d = 17.5$ in.

$h = 20$ in.

$d' = 2.5$ in.

$A_s = A'_s = 3$ in.^{2}

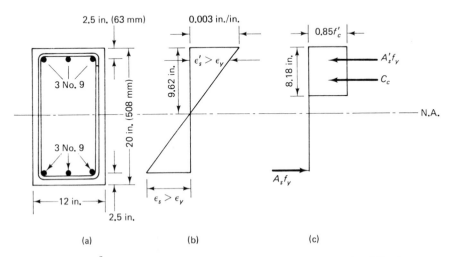

Figure 7.10 Column geometry—strain and stress diagrams (tension failure): (a) cross section; (b) strains; (c) stresses.

$$f'_c = 4000 \text{ psi}$$

$$f_y = 60{,}000 \text{ psi}$$

Solution

Using the results of Ex. 7.3, $e_b = 13.1$ in. $< e = 14$ in. Therefore failure will occur by the initial yielding of tension steel.

$$\rho = \rho' = \frac{A_s}{bd} = \frac{3}{12 \times 17.5} = 0.0143$$

$$m = \frac{60{,}000}{0.85 \times 4000} = 17.65$$

$$\frac{h - 2e}{2d} \quad \text{or} \quad 1 - \frac{e'}{d} = \frac{20 - 2 \times 14}{2 \times 17.5} = -0.2286$$

$$1 - \frac{d'}{d} = 1 - \frac{2.5}{17.5} = 0.8571$$

Using Eq. 7.19 or 7.20, we have

$$P_n = 0.85 \times 4000 \times 12 \times 17.5$$
$$\times \, [-0.2286 + \sqrt{(-0.2286)^2 + 2 \times 17.65 \times 0.0143 \times 0.8571}] = 333{,}979 \text{ lb}$$

$$a = \frac{P_n}{0.85 f'_c b} = \frac{333{,}979}{0.85 \times 4000 \times 12} = 8.19 \text{ in.}$$

$$c = \frac{8.19}{0.85} = 9.63 \text{ in.}$$

$$f'_s = 0.003 \times 29 \times 10^6 \left(\frac{9.63 - 2.5}{9.63} \right)$$

$$= 64{,}414 \text{ psi} > f_y \qquad \text{Therefore } f'_s = f_y \qquad \text{O.K.}$$

$$P_n = 333{,}438 \text{ lb } (1500.47 \text{ kN}) \text{ at } e = 14 \text{ in. } (356.6 \text{ mm})$$

(If the area of displaced concrete is taken into account, $P_n = 328{,}970$ lb.) If f'_s is less than f_y, a trial-and-adjustment procedure must be used for the analysis.

It should be emphasized that in each analysis (design) problem the balanced P_{nb}, M_{nb}, and hence e_b need to be evaluated to verify whether the appropriate expressions for tension failure or compression failure are applied in the solution.

7.5.5 Example 7.5: Analysis of a Column Controlled by Tension Failure; Stress in Compression Steel Less Than Yield Strength

A short rectangular reinforced concrete column is 12 in. × 15 in. (305 mm × 381 mm) as shown in Fig. 7.11 and is subjected to a load eccentricity $e = 12$ in. (305 mm). Calculate the safe nominal load strength P_n and the nominal moment strength M_n of the

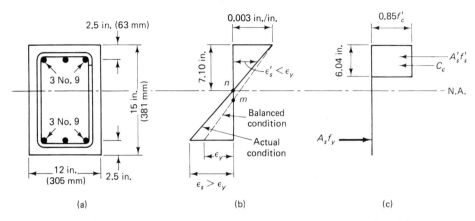

Figure 7.11 Column geometry—strain and stress diagrams (tension failure $f'_s < f_y$): (a) cross section; (b) strains; (c) stresses.

column section. Given:

$f'_c = 4000$ psi (27.6 MPa), normalweight concrete

$f_y = 60,000$ psi (414 MPa)

Eccentricity $e = 12$ in. (305 mm), and three No. 9 bars (28.7 mm diameter) for each of the compression and tension reinforcement.

Solution

To determine whether the failure is by crushing of concrete or yielding of steel, P_{nb} and e_b should be calculated first.

$$A_s = A'_s = 3 \text{ in.}^2 \ (19.4 \text{ cm}^2)$$

$$d = 12.5 \text{ in.}$$

$$c_b = \left(\frac{87,000}{87,000 + 60,000}\right)12.5 = 7.4 \text{ in.}$$

$$a_b = 0.85 \times 7.4 = 6.3 \text{ in.}$$

$$f'_s = 29 \times 10^6 \times 0.003\left(\frac{7.4 - 2.5}{7.4}\right) = 57,609 \text{ psi} < f_y$$

So the compression steel did not yield.

$$P_{nb} = 0.85 \times 4000 \times 12 \times 6.3 + 3 \times 57,609 - 3 \times 60,000 = 249,867 \text{ lb}$$

$$M_{nb} = 0.85 \times 4000 \times 12 \times 6.3\left(\frac{15}{2} - \frac{6.3}{2}\right) + 3 \times 57,609\left(\frac{15}{2} - 2.5\right)$$

$$+ 3 \times 60,000\left(12.5 - \frac{15}{2}\right) = 2,882,259 \text{ in.-lb}$$

$$e_b = \frac{M_{nb}}{P_{nb}} = \frac{2,882,259}{249,867} = 11.5 \text{ in.}$$

The specified eccentricity $e = 12$ in. is greater than e_b. Thus failure will occur by initial yielding of steel. Point m on the vertical axis of the strain diagram in Fig. 7.11b denotes the position of the neutral axis at the balanced condition. For this condition, calculations showed that the strain in the compression steel is less than its yield strain. As the neutral axis position rises to point n in the case of initial failure by yielding of the tension steel, the strain ϵ_s' in the compression steel will be less than that of the balanced condition, hence less than ϵ_y. Therefore the trial-and-adjustment method should be used for the calculation of P_n. Because $c_b = 7.4$ in., assume a slightly smaller depth to the neutral axis for initial tension failure. Try $c = 7.10$ in.

$$a = 0.85 \times 7.10 = 6.04 \text{ in.}$$

$$f_s' = 29 \times 10^6 \times 0.003 \left(\frac{7.10 - 2.5}{7.10} \right) = 56,366 \text{ psi}$$

Since failure is by yielding of tension steel, the stress in the tension steel is equal to the yield strength, or

$$f_s = f_y = 60,000 \text{ psi}$$

$$P_n = 0.85 \times 4000 \times 12 \times 6.04 + 3 \times 56,366 - 3 \times 60,000$$

$$= 235,531 \text{ lb}$$

$$M_n = 0.85 \times 4000 \times 12 \times 6.04(7.5 - 3.02) + 3 \times 56,366 \times 5 + 3$$

$$\times 60,000 \times 5$$

$$= 2,849,508 \text{ in.-lb (322 kN-m)}$$

$$e = \frac{M_n}{P_n} = 12.10 \text{ in. (305 mm)}$$

Therefore

$$P_n \simeq 236,000 \text{ lb (1050 kN)}$$

(If displaced concrete is taken into account in the calculations, $P_n = 235,050$ lb.)

7.5.6 Compression Failure in Rectangular Column Sections

For initial crushing of the concrete, the eccentricity e must be less than the balanced eccentricity e_b and the stress in the tensile reinforcement below yield—that is, $f_s < f_y$.

The analysis (design) process necessitates applying the basic equilibrium equations 7.6 and 7.7, using the trial-and-adjustment procedure and ensuring strain-compatibility checks at all stages. The procedure is summarized in Section 7.4.2 and the following example illustrates its use in the analysis and design of reinforced concrete columns.

7.5.7 Example 7.6: Analysis of a Column Controlled by Compression Failure; Trial-and-Adjustment Procedure

Calculate the nominal load P_n of the section in Ex. 7.1 (see Fig. 7.12) if the column is subjected to a load eccentricity $e = 10$ in. (254 mm). Given:

$b = 12$ in. (305 mm)

$d = 17.5$ in. (445 mm)

$h = 20$ in. (508 mm)

$d' = 2.5$ in.

$A_s = A_s' = 3.0$ in.^{2} (1940 mm^2)

$f_c' = 4000$ psi (27.6 MPa)

$f_y = 60,000$ psi (414 MPa)

Solution

Using the results of Ex. 7.3, eccentricity for balanced failure, $e_b = 13.1$ in., which is larger than the given eccentricity of 10 in. Therefore failure will occur by initial crushing of concrete at the compression face.

Trial 1

Assume that

$$c = 11.0 \text{ in. (254 mm)}$$

$$a = \beta_1 c = 0.85 \times 11.0 = 9.35 \text{ in.}$$

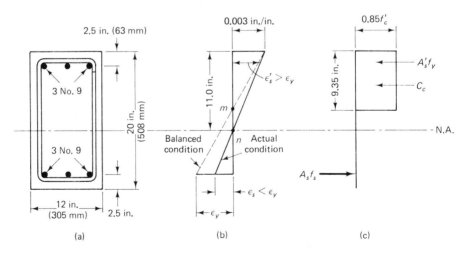

Figure 7.12 Column geometry—strain and stress diagrams (compression failure): (a) cross section; (b) strains; (c) stresses.

Using Eq. 7.8 gives

$$f_s' = 29 \times 10^6 \times 0.003\left(\frac{11.0 - 2.5}{11.0}\right) = 67,227 \text{ psi} > f_y$$

Therefore

$$f_s' = f_y = 60,000 \text{ psi}$$

Then using Eq. 7.9 yields

$$f_s = 29 \times 10^6 \times 0.003\left(\frac{17.5 - 11.0}{11.0}\right) = 51,409 \text{ psi}$$

And using Eq. 7.6, we have

$$P_n = 0.85 \times 4000 \times 12 \times 9.35 + 3 \times 60,000 - 3 \times 51,409$$

$$= 407,253 \text{ lb}$$

Finally, using Eq. 7.7, we obtain

$$M_n = 0.85 \times 4000 \times 12 \times 9.35\left(10 - \frac{9.35}{2}\right) + 3 \times 60,000(10 - 2.5)$$

$$+ 3 \times 51,409(17.5 - 10) = 4,538,083 \text{ in.-lb}$$

Photo 54 Compression side of eccentrically loaded column at failure. (Tests by Nawy, et al.)

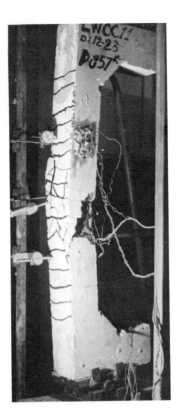

Photo 55 Tension side of eccentrically loaded column at rupture and cover spalling. (Tests by Nawy, et al.)

$$e = \frac{M_n}{P_n} = 11.14 \text{ in. } > 10 \text{ in.}$$

Therefore an axial load of 407,253 lb can be applied at an eccentricity of 11.14 in.

Trial 2

Assume that $c = 11.5$ in.; so $a = 0.85 \times 11.5 = 9.77$ in.

$$f'_s = 60,000 \text{ psi}$$

$$f_s = 45,391 \text{ psi}$$

$$P_n = 442,647 \text{ lb}$$

$$M_n = 4,410,218 \text{ in.-lb}$$

$$e = 9.96 \text{ in. } \simeq \text{ given eccentricity of 10 in.}$$

Therefore, for $e = 10$ in. (254 mm), P_n can be assumed to be 442,647 lb. (If displaced concrete is taken into account in the calculations, $P_n = 432,445$ lb.)

7.6 WHITNEY'S APPROXIMATE SOLUTION

Empirical expressions proposed by Whitney can rapidly be used in lieu of the trial-and-adjustment method, although with some loss in accuracy.

Whitney's solution is based on the following assumptions.

1. Reinforcement is symmetrically placed in single layers parallel to the axis of bending in rectangular sections.
2. Compression steel has yielded.
3. Concrete displaced by the compression steel is negligible compared to the total concrete area in compression; so no correction is made for the concrete displaced by the compression steel.
4. In order to calculate the contribution C_c of the concrete, the depth of the stress block is assumed to be $0.54d$, corresponding to an average value of a for balanced conditions in rectangular sections.
5. The interaction curve in the compression zone is a straight line, as shown in Fig. 7.14.

For most cases, Whitney's method leads to a conservative solution except when the factored load P_u has a value higher than the balanced load P_{ub} and the external eccentricity e is very small. Otherwise the method leads to a nonconservative solution as shown in the shaded area BST of Fig. 7.14.

If compression controls, the equation can be written

$$P_n = \frac{A_s' f_y}{[e/(d - d')] + 0.5} + \frac{bhf_c'}{(3he/d^2) + 1.18} \tag{7.23}$$

The following example illustrates the use of this equation.

7.6.1 Example 7.7: Analysis of a Column Controlled by Compression Failure; Whitney's Equation

Calculate the nominal strength load P_n for the section in Ex. 7.6, using Whitney's equation if the load eccentricity is (a) $e = 6$ in. (152.4 mm) and (b) $e = 10$ in. (254 mm).

Solution

(a) $e = 6$ in.:

$$P_n = \frac{3 \times 60,000}{[6/(17.5 - 2.5)] + 0.5} + \frac{12 \times 20 \times 4000}{[(3 \times 20 \times 6)/17.5^2] + 1.18}$$

$$= 607,555 \text{ lb } (2734 \text{ kN})$$

The exact solution, using trial and adjustment and including the displaced concrete, gives $P_n = 608,458$ lb (2738 kN). The approximate solution is conservative.

(b) $e = 10$ in.: Using Eq. 7.23,

$$P_n = \frac{3 \times 60{,}000}{[10/(17.5 - 2.5)] + 0.5} + \frac{12 \times 20 \times 4000}{[(3 \times 20 \times 10)/17.5^2] + 1.18}$$

$$= 460{,}098 \text{ lb } (2070.4 \text{ kN})$$

The exact solution, using the trial-and-adjustment procedure and including the effect of the displaced concrete, gives $P_n = 433{,}138$ lb (1960 kN), showing that the approximate solution is not conservative, as discussed above.

7.7 COLUMN STRENGTH REDUCTION FACTOR ϕ

Failure is initiated in members subject to flexure and relatively small axial loads by yielding of the tension reinforcement and takes place in an increasingly ductile manner. Thus for small axial loads, it is reasonable to permit an increase in the ϕ factor from that required for pure compression members. When the axial load vanishes, the member is subjected to pure flexure and the strength reduction factor ϕ becomes 0.90. Figure 7.13a and b shows the zone in which the value of ϕ can be increased from 0.7 to 0.9 for tied columns and 0.75 to 0.9 for spiral columns. As the factored design compression load ϕP_n decreases beyond $0.1 A_g f_c'$, in Fig. 7.13a, the ϕ factor is increased from 0.7 to 0.9 for tied columns and 0.75 to 0.9 for spiral columns. For those cases where the value of P_{nb} is less than $0.1 A_g f_c'$, ϕ values are increased when the load $P_u < P_{ub}$ or $\phi P_n < \phi P_{nb}$, as in Fig. 7.13b.

The value $0.10 f_c' A_g$ is chosen by the ACI code as the design axial load value ϕP_n below which the ϕ factor could safely be increased for most compression members. In summary, if the initial failure is in compression, the strength reduction factor ϕ is always 0.70 for tied columns and 0.75 for spirally reinforced columns.

The following expressions give variations in the value of ϕ for symmetrically reinforced compression members. The columns should have an effective depth not less than 70% of the total depth and the steel reinforcement should have a yield strength not exceeding 60,000 psi. For tied columns,

$$\phi = 0.90 - \frac{0.20\phi P_n}{0.1 f_c' A_g} \geq 0.70 \tag{7.24}$$

For spirally reinforced columns,

$$\phi = 0.90 - \frac{0.15\phi P_n}{0.1 f_c' A_g} \geq 0.75 \tag{7.25}$$

where $$P_u = \phi P_n \tag{7.26}$$

In both Eqs. 7.24 and 7.25, if ϕP_{nb} is less than $0.1 f_c' A_g$, then ϕP_{nb} should be substituted for $0.1 f_c' A_g$ in the denominator, using $0.70 P_{nb}$ for tied and $0.75 P_{nb}$ for spirally reinforced columns. Figure 7.13c presents the graphical representation of Eqs. 7.24 and 7.25.

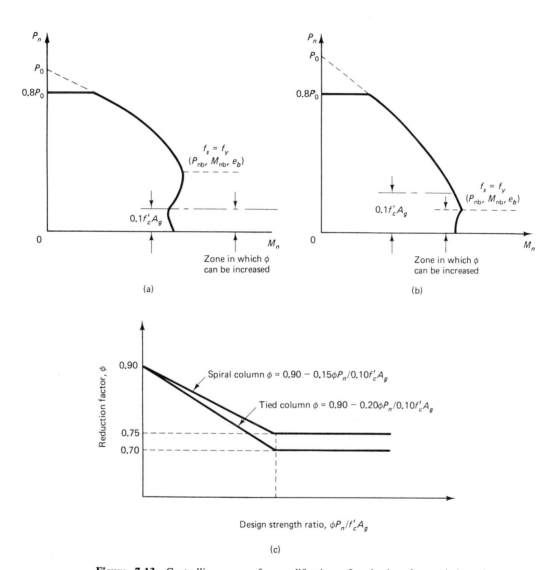

Figure 7.13 Controlling zones for modification of reduction factor ϕ in columns: (a) $0.1f'_c A_g < \phi P_{nb}$; (b) $0.1f'_c A_g > \phi P_{nb}$; (c) variation of ϕ for symmetrically reinforced compression members (P_n is the nominal axial strength at given eccentricity).

7.7.1 Example 7.8: Calculation of Design Load Strength ϕP_n from Nominal Resisting Load P_n

Calculate the design loads P_u in Exs. 7.1 to 7.7, using the appropriate ϕ reduction factors.

Solution

Example 7.1:

$$P_{n(\text{max})} = 924{,}480 \text{ lb, tied column}$$

Therefore

$$\phi = 0.7.$$

$$\phi P_{0(\text{max})} = 0.7 \times 924{,}480 = 647{,}136 \text{ lb}$$

Example 7.2:

$$P_{n(\text{max})} = 1{,}135{,}501 \text{ lb, spiral column}$$

Therefore

$$\phi = 0.75.$$

$$\phi P_{0(\text{max})} = 0.75 \times 1{,}135{,}501 = 851{,}626 \text{ lb}$$

Example 7.3:

$$P_{nb} = 359{,}448 \text{ lb, tied column}$$

Therefore

$$\phi = 0.7.$$

$$\phi P_{nb} = 251{,}614 \text{ lb}$$

$$e_b = 13.1 \text{ in.}$$

Example 7.4:

$$P_n = 333{,}438 \text{ lb, tied column}$$

Failure is by yielding of steel at the tension side. Therefore it has to be checked if the ϕ value is larger than 0.7.

$$0.1 A_g f'_c = 0.1 \times 12 \times 20 \times 4000 = 96{,}000 \text{ lb} < 0.7 P_n = 233{,}407 \text{ lb}$$

Therefore

$$\phi = 0.7.$$

$$\phi P_n = 0.7 \times 333{,}438 = 233{,}407 \text{ lb}$$

Example 7.5:

$$\phi = 0.7 \text{ (similar to previous case)}$$

$$\phi P_n = 0.7 \times 236{,}200 = 165{,}340 \text{ lb}$$

Example 7.6:

$$P_n = 440{,}600 \text{ lb, tied column}$$

Failure is by crushing of concrete, axial force greater than balanced load. Therefore

$$\phi = 0.7$$

$$\phi P_n = 0.7 \times 442{,}647 = 309{,}853 \text{ lb}$$

Example 7.7:
(a) P_n = 460,000 lb, tied column. Failure is by crushing of concrete. Therefore

$$\phi = 0.7$$

$$\phi P_n = 0.7 \times 607,555 = 425,288 \text{ lb}$$

(b) P_n = 460,098 lb

$$\phi P_n = 0.7 \times 460,098 = 322,070 \text{ lb}$$

7.8 LOAD–MOMENT STRENGTH INTERACTION DIAGRAMS (P–M DIAGRAMS) FOR COLUMNS CONTROLLED BY MATERIAL FAILURE

From the discussion in Section 7.3 and 7.4 and the numerical examples presented one can postulate that the capacity of reinforced concrete sections to resist combined axial and bending loads can be expressed by *P–M* interaction diagrams to relate the axial load to the bending moment in compression members. Figure 7.14 presents one such

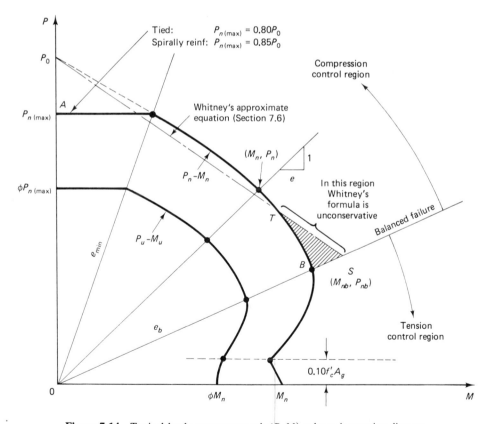

Figure 7.14   Typical load-moment strength (*P–M*) column interaction diagram.

diagram. Notice that the approximation made in using the Whitney empirical approach is not always conservative, particularly when the factored P_u is close to the balanced case.

Each point on the curve represents one combination of nominal load strength P_n and nominal moment strength M_n corresponding to a particular neutral axis location. The interaction diagram is separated into the tension control region and the compression control region by the balanced condition at point B. The following example illustrates the construction of the P–M diagram for a typical rectangular section.

7.8.1 Example 7.9: Construction of a Load–Moment Interaction Diagram

Construct a P–M diagram for a rectangular column (see Fig. 7.15) having the following geometry: width $b = 12$ in. (305 mm), thickness $h = 14$ in. (356 mm), reinforcement: four No. 11 bars (35.8 mm diameter). Given:

$$f_c' = 6000 \text{ psi (41.4 MPa)}$$

$$f_y = 60{,}000 \text{ psi (414 MPa)}$$

$$d' = 3.0 \text{ in. (76.2 mm)}$$

Solution

Concentric Load

$$A_s = A_s' = 3.12 \text{ in.}^2$$

$$P_{n(\text{max})} = 0.80(0.85f_c'A_g + A_{st}f_y)$$

$$= 0.80(0.85 \times 6000 \times 14 \times 12 + 2 \times 3.12 \times 60{,}000)$$

$$= 984{,}960 \text{ lb}$$

$$\phi P_{n(\text{max})} = 0.7P_{n(\text{max})} = 689{,}472 \text{ lb}$$

(If displaced concrete is taken into account in the calculations, $P_{n(\text{max})} = 959{,}500$ lb and $\phi P_{n(\text{max})} = 671{,}650$ lb.)

Balanced condition

$$d = 14.0 - 3.0 = 11.0 \text{ in.}$$

$$c_b = \frac{87{,}000}{87{,}000 + 60{,}000} \times 11.0 = 6.51 \text{ in.}$$

$$\epsilon_s' = 0.003 \times \frac{6.51 - 3.0}{6.51} = 0.0016 \text{ in./in.}$$

$$< \frac{f_y}{E_s} = \frac{60}{29{,}000} = 0.00207$$

$$f_s' = 29 \times 10^6 \times 0.0016 = 46{,}400 \text{ psi}$$

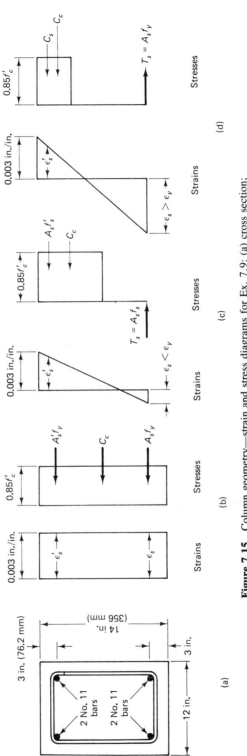

Figure 7.15 Column geometry—strain and stress diagrams for Ex. 7.9: (a) cross section; (b) concentric load (compression failure); (c) compression failure; (d) tension failure.

225

$$\beta_1 = 0.85 - \frac{0.05(6000 - 4000)}{1000} = 0.75 \text{ in.}$$

$$a_b = \beta_1 c_b = 0.75 \times 6.51 = 4.88 \text{ in.}$$

$$P_{nb} = 0.85 f_c' b a_b + A_s' f_s' - A_s f_y$$

or

$$P_{nb} = 0.85 \times 6000 \times 12 \times 4.88 + 3.12 \times 46,400 - 3.12 \times 60,000$$

$$= 298,656 + 144,768 - 187,200 = 256,224 \text{ lb}$$

$$M_{nb} = 0.85 f_c' b a_b \left(\bar{y} - \frac{a}{2}\right) + A_s' f_s' (\bar{y} - d') + A_s f_y (d - \bar{y})$$

$\bar{y}$ = distance from extreme compression fibers to the plastic centroid

$$= \frac{h}{2} = \frac{14.0}{2} = 7 \text{ in.}$$

or

$$M_{nb} = 298,656\left(7 - \frac{4.88}{2}\right) + 144,768(7 - 3) + 187,200(11 - 7)$$

$$= 1,361,871 + 579,072 + 748,800 = 2,689,743 \text{ in.-lb}$$

$$e_b = \frac{M_{nb}}{P_{nb}} = \frac{2,689,743}{256,224} = 10.50 \text{ in.}$$

$$\phi P_{nb} = 0.7 P_{nb} = 179,357 \text{ lb} \qquad \phi M_{nb} = 1,882,820 \text{ in.-lb}$$

$$e_b = 10.5 \text{ in.}$$

(If displaced concrete is considered in the calculations, $\phi P_{nb} = 169,444$ lb and $e_b = 10.9$ in.)

Pure bending M_{n0}

Neglect the contribution of A_s' to the moment strength as insignificant when $P_u = 0$. Thus

$$a = \frac{A_s f_y}{0.85 f_c' b} = \frac{3.12 \times 60,000}{0.85(6000)(12)} = \frac{187,200}{61,200} = 3.06 \text{ in.}$$

$$c = \frac{a}{\beta_1} = \frac{3.06}{0.75} = 4.08 \text{ in.} \qquad \epsilon_s' = 0.00079 \qquad f_s' = 22,910 \text{ psi}$$

$$M_{n0} = A_s f_y \left(d - \frac{a}{2}\right) = 187,200(11 - 1.53) = 1,772,784 \text{ in.-lb}$$

$$\phi M_{n0} = 0.9 \times 1,772,784 = 1,595,506 \text{ in.-lb}$$

For $c = 10$ in. $> c_b$: compression controls

$$\epsilon_s' = 0.003 \times \frac{10 - 3}{10} = 0.0021 \text{ in./in.}$$

$$\epsilon_y = \frac{f_y}{E_s} = \frac{60,000}{29 \times 10^6} = 0.0021 \simeq \epsilon_s' \qquad \text{Therefore } f_s' = f_y$$

$$\epsilon_s = 0.003 \times \frac{11 - 10}{10} = 0.0003$$

$$f_s = 0.0003 \times 29 \times 10^6 = 8700 \text{ psi}$$

$$a = \beta_1 c = 0.75 \times 10 = 7.5 \text{ in.}$$

$$C_c = 0.85(6000)(12)(7.5) = 459,000 \text{ lb}$$

$$C_s = 3.12(60,000) = 187,200 \text{ lb}$$

$$T_s = 3.12(8700) = 27,144 \text{ lb}$$

$$P_n = C_c + C_s - T_s$$

or

$$P_n = (459,000) + (187,200 - 27,144) = 619,056 \text{ lb}$$

$$M_n = C_c\left(\bar{y} - \frac{a}{2}\right) + C_s(\bar{y} - d') + T_s(d - \bar{y})$$

$$= 459,000(7 - 3.75) + 187,200(7 - 3) + 27,144(11 - 7)$$

$$= 1,491,750 + 748,800 + 108,576 = 2,349,126$$

$$\phi P_n = 0.7 P_n = 0.7 \times 619,056 = 433,339 \text{ lb}$$

(With displaced concrete taken into account, $\phi P_n = 422,200$ lb and $e = 3.79$ in.)

For $c = 4.2$ in. $< c_b$: tension control

$$a = \beta_1 c = 0.75 \times 4.2 = 3.15 \text{ in.}$$

$$\epsilon_s' = 0.003 \times \frac{4.2 - 3}{4.2} = 0.0009 < \epsilon_y$$

$$f_s' = 0.0009 \times 29 \times 10^6 = 24,857 \text{ psi}$$

$$f_s = f_y = 60,000 \text{ psi}$$

$$C_c = 0.85(6000)(12)(3.15) = 192,780 \text{ lb}$$

$$C_s = 3.12(24,857) = 77,554 \text{ lb}$$

$$T_s = 3.12(60,000) = 187,200 \text{ lb}$$

$$P_n = C_c + C_s - T_s = 192,780 + 77,554 - 187,200 = 83,134 \text{ lb}$$

$$M_n = C_c\left(\bar{y} - \frac{a}{2}\right) + C_s(\bar{y} - d') + T_s(d - \bar{y})$$

or

$$M_n = 192,780\left(7 - \frac{3.15}{2}\right) + 77,554(7 - 3) + 187,200(11 - 7)$$

$$= 1{,}044{,}868 + 310{,}216 + 748{,}800 = 2{,}103{,}884 \text{ in.-lb}$$

$$e = \frac{M_n}{P_n} = \frac{2{,}103{,}884}{83{,}134} = 25.31 \text{ in.}$$

Assuming that $\phi = 0.70$,

$$\phi P_n = 0.70 \times 83{,}134 = 58{,}194 \text{ lb}$$

$$0.1 A_g f_c' = 0.10(12 \times 14) \times 6000 = 100{,}800 \text{ lb}$$

$$\phi P_n < 0.1 A_g f_c' \qquad \text{Hence } \phi > 0.70$$

Therefore

$$\phi = 0.90 - \frac{0.2 \times 58{,}194}{0.1 \times 1{,}008{,}000} = 0.792$$

$$\phi P_n = 0.792 \times 83{,}134 = 65{,}842$$

Verify the first trial ϕ value:

$$\phi = 0.90 - \frac{0.2 \times 65{,}842}{0.1 \times 1{,}008{,}000} = 0.78$$

$$\phi P_n = 0.78 \times 83{,}134 = 64{,}845 \text{ lb}$$

(If displaced concrete is considered, $\phi P_n = 47{,}056$ lb and $e = 30.4$ in.)

For $0.10 f_c' A_g = \phi P_n$

Assume by trial and adjustment a value of $c = 4.85$ in.

$$a = \beta_1 c = 0.75 \times 4.85 = 3.64 \text{ in.}$$

$$\epsilon_s' = 0.003 \times \frac{4.85 - 3}{4.85} = 0.00114 < \epsilon_y$$

$$f_s' = 0.00114 \times 29 \times 10^6 = 33{,}060 \text{ psi}$$

$$C_c = 0.85 \times 6000 \times 12 \times 3.64 = 227{,}768 \text{ lb}$$

$$C_s = 3.12 \times 33{,}060 = 103{,}147 \text{ lb}$$

$$T_s = 3.12 \times 60{,}000 = 187{,}200 \text{ lb}$$

$$P_n = C_c + C_s - T_s = 227{,}768 + 103{,}147 - 187{,}200 = 143{,}715 \text{ lb}$$

$$\phi P_n = 0.70 \times 143{,}715 = 100{,}619 \text{ lb}$$

$$0.1 f_c' A_g = 100{,}800 \text{ lb} \cong \phi P_n \qquad \text{assumed } c \text{ value O.K.}$$

$$M_n = C_c\left(\bar{y} - \frac{a}{2}\right) + C_s(\bar{y} - d') + T_s(d - \bar{y}) = 227{,}768\left(7 - \frac{3.64}{2}\right)$$

$$+ 103{,}147(7 - 3) + 187{,}200(11 - 7) = 2{,}341{,}226 \text{ lb}$$

$$e = \frac{M_n}{P_n} = \frac{2{,}341{,}226}{143{,}715} = 16.29 \text{ in.}$$

Additional points on the diagram are calculated by assigning other values for the neutral axis depth c. The interaction diagram is presented in Fig. 7.16. Sets of charts of nondimensional interaction diagrams are available for various codes and units to facilitate speedy analysis and design in engineering offices. A typical chart from the ACI 340 SP-17(81) handbook is shown in Fig. 7.17, and additional charts are given in Appendix C.

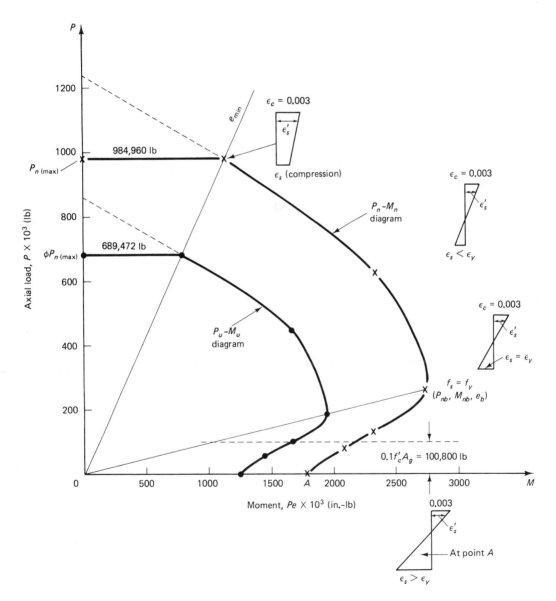

Figure 7.16 Interaction P–M diagram for Ex. 7.9.

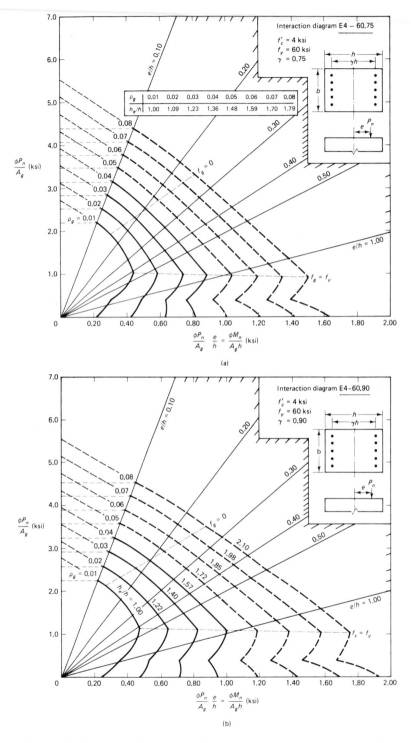

Figure 7.17 Typical nondimensional column interaction charts: (a) chart for small column sizes; (b) chart for larger column sizes.

7.9 PRACTICAL DESIGN CONSIDERATIONS

The following guidelines should be followed in the design and arrangement of reinforcement to arrive at a practical design.

7.9.1 Longitudinal or Main Reinforcement

Most columns are subjected to bending moment in addition to axial force. For this reason and to ensure some ductility, a minimum of 1% reinforcement should be provided in the columns. A reasonable reinforcement ratio is between 1.5 and 3.0%. Occasionaly in high-rise buildings where column loads are very large, 4% reinforcement is not unreasonable. Even though the code allows a maximum of 8% for longitudinal reinforcement in columns, it is not advisable to use more than 4% in order to avoid reinforcement congestion, especially at beam–column junctions.

A minimum of four longitudinal bars should be used in the case of tied columns. At least six longitudinal bars should be used for spiral columns to provide hoop action in the spirals; see the ACI code for further discussion.

7.9.2 Lateral Reinforcement for Columns

Lateral Ties

Lateral reinforcement is required to prevent spalling of the concrete cover or local buckling of the longitudinal bars. The lateral reinforcement could be in the form of ties evenly distributed along the height of the column at specified intervals. Longitudinal bars spaced more than 6 in. apart should be supported by lateral ties, as shown in Fig. 7.18.

The following guidelines are to be followed for the selection of the size and spacing of ties.

1. The size of the tie should not be less than a No. 3 (9.5 mm) bar. If the longitudinal bar size is larger than No. 10 (32 mm), then No. 4 (12 mm) bars at least should be used as ties.

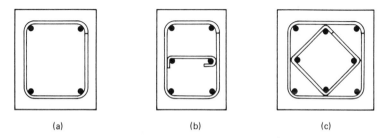

(a)   (b)   (c)

Figure 7.18 Typical ties arrangement for four, six, and eight longitudinal bars in a column: (a) one tie; (b) two ties; (c) two ties.

2. The vertical spacing of the ties must not exceed:
    (a) 48 times the diameter of the tie
    (b) 16 times the diameter of the longitudinal bar
    (c) Least lateral dimension of the column

Figure 7.18 shows a typical arrangement of ties for four, six, and eight longitudinal bars in a column cross section.

Spirals

The other type of lateral reinforcement is spirals or helical lateral reinforcement as shown in Fig. 7.19. They are particularly useful in increasing ductility or member toughness; consequently, they are mandatory in high-earthquake-risk regions. Normally concrete outside the confined core of the spirally reinforced column can totally spall under unusual and sudden lateral forces such as earthquake-induced forces. The columns must be able to sustain most of the load even after the spalling of the cover in order to prevent the collapse of the building. So the spacing and size of spirals are designed to maintain most of the load-carrying capacity of the column, even under such severe load conditions.

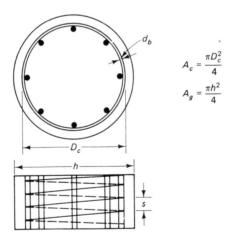

$$A_c = \frac{\pi D_c^2}{4}$$

$$A_g = \frac{\pi h^2}{4}$$

Figure 7.19 Helical or spiral reinforcement for columns.

Closely spaced spiral reinforcement increases the ultimate load capacity of columns. The spacing or pitch of the spiral is so chosen that the load capacity due to the confining spiral action compensates for the loss due to spalling of the concrete cover.

Equating the increase in strength due to confinement and the loss of capacity in spalling and incorporating a safety factor of 1.2, the following minimum spiral reinforcement ratio ρ_s is obtained:

$$\rho_s = 0.45\left(\frac{A_g}{A_c} - 1\right)\frac{f_c'}{f_{sy}} \qquad (7.27)$$

where $\rho_s = \dfrac{\text{volume of the spiral steel per one revolution}}{\text{volume of concrete core contained in one revolution}}$

$$A_c = \frac{\pi D_c^2}{4} \tag{7.28a}$$

$$A_g = \frac{\pi h^2}{4} \tag{7.28b}$$

h = diameter of the column
a_s = cross-sectional area of the spiral
d_b = nominal diameter of the spiral wire
D_c = diameter of the concrete core out-to-out of the spiral
f_{sy} = yield strength of the spiral reinforcement

To determine the pitch s of the spiral, calculate ρ_s using Eq. 7.27, choose a bar diameter d_b for the spiral, and calculate a_s; then obtain pitch s, using Eq. 7.30b.

The spiral reinforcement ratio ρ_s, can be written

$$\rho_s = \frac{a_s \pi (D_c - d_b)}{(\pi/4)D_c^2 s} \tag{7.29}$$

Therefore

$$\text{Pitch } s = \frac{a_s \pi (D_c - d_b)}{(\pi/4)D_c^2 \rho_s} \tag{7.30a}$$

or

$$s = \frac{4a_s(D_c - d_b)}{D_c^2 \rho_s} \tag{7.30b}$$

The spacing or pitch of spirals is limited to a range of 1 to 3 in. (25.4 to 76.2 mm) and the diameter should be at least $\frac{3}{8}$ in. (9.53 mm). The spirals should be well anchored by providing at least $1\frac{1}{2}$ extra turns when splicing of spirals rather than welding is used.

7.10 OPERATIONAL PROCEDURE FOR THE DESIGN OF NONSLENDER COLUMNS

The following steps can be used for the design of nonslender (short) columns where the behavior is controlled by material failure.

1. Evaluate the factored external axial load P_u and factored moment M_u. Calculate the eccentricity $e = M_u/P_u$.
2. Assume a cross section and the type of vertical reinforcement to be used. Fractional dimensions are to be avoided in selecting column sizes.
3. Assume a reinforcement ratio ρ between 1 and 4% and obtain the reinforcement area.

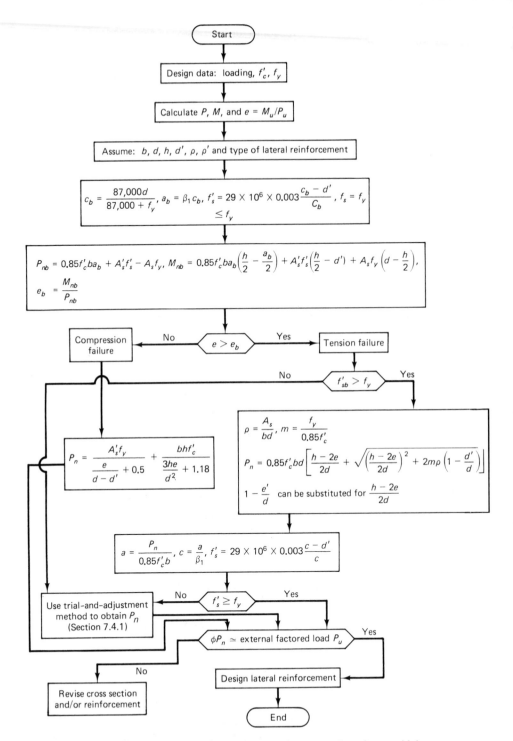

Figure 7.20 Flowchart for design of nonslender rectangular columns with bars on two faces only.

4. Calculate P_{nb} for the assumed section and determine the type of failure, whether by initial yielding of the steel or initial crushing of the concrete.
5. Check for the adequacy of the assumed section. If the section cannot support the factored load or it is oversized, hence uneconomical, revise the cross section and (or) the reinforcement and repeat steps 4 and 5.
6. Design the lateral reinforcement.

Figure 7.20 presents a flowchart for the sequence of calculations.

7.11 NUMERICAL EXAMPLES FOR ANALYSIS AND DESIGN OF NONSLENDER COLUMNS

7.11.1 Example 7.10: Design of a Column with Large Eccentricity; Initial Tension Failure

The tied reinforced concrete column in Fig. 7.21 is subjected to a service axial force due to dead load = 65,000 lb (289 kN) and a service axial force due to live load = 125,000 lb (556 kN). Eccentricity to the plastic centroid is e = 16 in. (406 mm).

Design the longitudinal and lateral reinforcement for this column, assuming a nonslender column with a total reinforcement ratio between 2 and 3%. Given:

f_c' = 4000 psi (27.6 MPa), normalweight concrete

f_y = 60,000 psi (414 MPa)

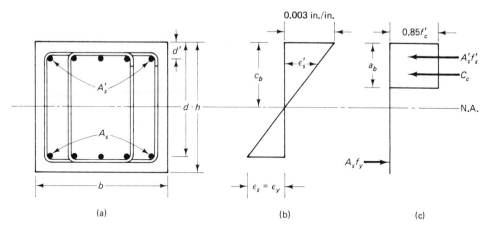

Figure 7.21 Column geometry—strain and stress diagrams in Ex. 7.10 (balanced failure): (a) cross section; (b) strains; (c) stresses.

Solution

Calculate the factored external load and moment (Step 1)

$P_u = 1.4D + 1.7L = 1.4 \times 65,000 + 1.7 \times 125,000 = 303,500$ lb (1350 kN)

$P_u e = 303,500 \times 16 = 4,856,000$ in.-lb (549 kN-m)

Assume a section 20 in. × 20 in. and a total reinforcement ratio of 3% (Steps 2 and 3)

Assume that $\rho = \rho' = A_s/bd = 0.015$ and $d' = 2.5$ in.

$$A_s = A'_s = 0.015 \times 20(20 - 2.5) = 5.25 \text{ in.}^2$$

Try five No. 9 bars—5.00 in.² on each face (3225 mm²)

$$\rho = \frac{5.00}{20 \times 17.5} = 0.0143$$

Check whether the given factored axial load P_u is greater than the balanced load, ϕP_{nb} (Step 4)

$$c_b = d\frac{87,000}{87,000 + f_y} = 17.5\left(\frac{87,000}{87,000 + 60,000}\right) = 10.4 \text{ in.}$$

$$a_b = \beta_1 c_b = 0.85 \times 10.4 = 8.82 \text{ in.}$$

$$f'_s = 0.003 \times 29 \times 10^6\left(\frac{10.4 - 2.5}{10.4}\right)$$

$$= 66,086 \text{ psi} > f_y$$

Therefore use $f'_s = f_y$. Using Eq. 7.6 gives

$$P_{nb} = 0.85f'_c ba_b + A'_s f_y - A_s f_y$$

$$= 0.85 \times 4000 \times 20 \times 8.82 = 599,760 \text{ lb (2670 kN)}$$

$$\phi P_{nb} = 0.7 \times P_{nb} = 419,832 \text{ lb}$$

Because the given load $P_u = 303,500$ lb is less than ϕP_{nb}, the column will fail by initial yielding of the tension reinforcement.

Check the adequacy of the section (Step 5)

Using Eq. 7.19 or 7.20 yields

$$\rho = 0.0143 \qquad m = \frac{60,000}{0.85 \times 4000} = 17.65$$

$$\frac{h - 2e}{2d} \quad \text{or} \quad 1 - \frac{e'}{d} = \frac{20 - 32}{2 \times 17.5} = -0.34$$

$$1 - \frac{d'}{d} = 1 - \frac{2.5}{17.5} = 0.857$$

$$P_n = 0.85 \times 4000 \times 20 \times 17.5(-0.34$$

$$+ \sqrt{0.12 + 2 \times 0.0143 \times 17.65 \times 0.857})$$

$$= 480,015 \text{ lb}$$

$$\phi P_n = 0.7 \times 480{,}015 = 336{,}011 \text{ lb } (1512 \text{ kN})$$

$$\phi P_n > 0.1 A_g f_c' \qquad \text{Therefore } \phi = 0.7 \qquad \text{O.K.}$$

Check if the compression steel stress $f_s' = f_y$:

$$a = \frac{480{,}015}{0.85 \times 4000 \times 20} = 7.06 \text{ in.}$$

$$c = \frac{a}{\beta_1} = 8.3 \text{ in.}$$

$$f_s' = 0.003 \times 29 \times 10^6 \left(\frac{8.3 - 2.5}{8.3} \right) = 60{,}795 \text{ psi} > f_y$$

Therefore $f_s' = f_y$. O.K.

An external load of 303,500 lb is less than 336,011 lb. Thus the design is satisfactory. (Using more exact computer programs, $\phi P_n = 327{,}964$ lb, which shows that using the preceding results can sometimes be unconservative.) Therefore adopt a section 20 in. × 20 in. (508 mm × 508 mm) with five No. 9 bars on each side (5 bars 28.6 mm diameter) having $d = 17.5$ in. (445 mm).

Design of ties (Step 6)

Using No. 3 ties, spacing will be the minimum of

$16 \times \frac{9}{8} = 18$ in.

$48 \times \frac{3}{8} = 18$ in.

Least dimension of 20 in.

Therefore provide No. 3 ties at 18 in. center to center (9.53 mm diameter at 457 mm center to center).

7.11.2 Example 7.11: Design of a Column with Small Eccentricity; Initial Compression Failure

A nonslender column shown in Fig. 7.22 is subjected to a factored $P_u = 365{,}000$ lb (1620 kN) and a factored $M_u = 1{,}640{,}000$ in.-lb (185 kNm). Assume that the gross reinforcement ratio $\rho_g = 1.5$ to 2% and that the effective cover to the center of the longitudinal steel is $d' = 2\frac{1}{2}$ in. (63.5 mm). Design the column section and the necessary longitudinal and transverse reinforcement. Given:

$f_c' = 4500$ psi (31.03 MPa), normalweight concrete

$f_y = 60{,}000$ psi (414 MPa)

Solution

Calculation of factored design loads (Step 1)

$$P_u = 365{,}000 \text{ lb}$$

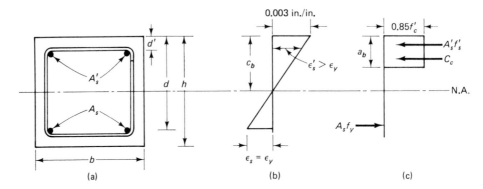

Figure 7.22 Column geometry—strain and stress diagrams in Ex. 7.11: (a) cross section; (b) strains (balanced case); (c) stresses.

$$e = \frac{1,640,000}{365,000} = 4.5 \text{ in. (114 mm)}$$

Assume a 15 in. × 15 in. (d = 12.5 in.) section (Steps 2 and 3)

Assume that the reinforcement ratio $\rho = \rho' = 0.01$.

$$A_s = A_s' \approx 0.01 \times 15 \times 12.5 = 1.875 \text{ in.}^2$$

Provide two No. 9 bars on each side.

$$A_s = A_s' = 2.0 \text{ in.}^2 \text{ (1290 mm}^2\text{)}$$

Check whether the given load is less or greater than P_{ub} (Step 4)

$$d = 15.0 - 2.5 = 12.5 \text{ in. (317.5 mm)}$$

$$c_b = \frac{87,000}{87,000 + 60,000} \times 12.5 = 7.4 \text{ in.}$$

$$\beta_1 = 0.85 - 0.05\left(\frac{4500 - 4000}{1000}\right) = 0.825$$

$$a = \beta_1 c = 0.825 \times 7.4 = 6.11 \text{ in.}$$

$$\epsilon_s' = 0.003 \times \frac{7.4 - 2.5}{7.4}$$

$$= 0.00199 \text{ in./in.} < \frac{f_y}{E_s}$$

$$f_s' = E_s \epsilon_s' = 29,000 \times 10^3 \times 0.00199 = 57,608 \text{ psi}$$

$$\phi P_{nb} = 0.7(0.85 f_c' b a_b + A_s' f_s' - A_s f_y)$$

$$= 0.7(0.85 \times 4500 \times 15 \times 6.11 + 2 \times 57,610 - 2 \times 60,000)$$

$$= 242,050 \text{ lb } (1080 \text{ kN})$$

$\phi P_{nb} < P_u$ compression failure controls

Check the adequacy of the section (Step 5)

$$P_n = \frac{A_s' f_y}{\dfrac{e}{d - d'} + 0.5} + \frac{bh f_c'}{\dfrac{3he}{d^2} + 1.18}$$

$$= \frac{2 \times 60,000}{\dfrac{4.5}{12.5 - 2.5} + 0.5} + \frac{15 \times 15 \times 4500}{\dfrac{3 \times 15 \times 4.5}{12.5^2} + 1.18}$$

$$= 535,200 \text{ lb } (2380 \text{ kN})$$

$$= \phi P_n = 0.7 \times 535,200 = 374,000 \text{ lb} > 365,000 \text{ kips}$$

Therefore the section is adequate to carry the load. Results from more exact computer solutions give $\phi P_n = 372,247$ lb. Use a column 15 in. $\times$ 15 in. ($d = 12.5$ in.) with two No. 9 bars on each face (381 mm $\times$ 381 mm size with two bars 28.6 mm diameter on each face).

Design of ties (Step 6)

Using No. 3 ties, the spacing will be the minimum of:

$16 \times \frac{9}{8} = 18$ in.

$48 \times \frac{3}{8} = 18$ in.

15 in.

Therefore provide No. 3 ties at 15 in. (9.53 mm diameter at 381 mm center to center).

SELECTED REFERENCES

7.1 Whitney, C. S., "Plastic Theory of Reinforced Concrete Design," *Transactions of the ASCE,* Vol. 107, 1942, pp. 251–326.

7.2 American Concrete Institute, *Design Handbook in Accordance with the Strength Design Method,* Vol. 2; *Columns,* Publications S-17A (78), ACI, Detroit, 1978, 191 pp.

7.3 ACI Committee 318, "Building Code Requirements for Reinforced Concrete, ACI Standard 318-83," American Concrete Institute, Detroit, 1983, 111 pp., and the "Commentary on Building Code Requirements for Reinforced Concrete," 1983, 155 pp.

PROBLEMS FOR SOLUTION

7.1 Calculate the axial load strength P_n for columns having the cross sections shown in Fig. 7.23. Assume zero eccentricity for all cases. Cases (a), (b), and (c) are tied columns.

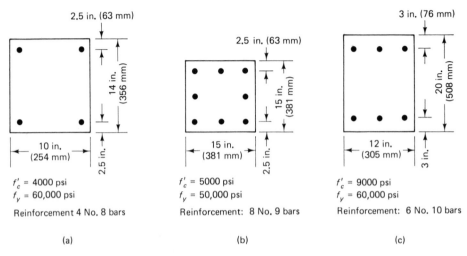

Figure 7.23 Column sections.

7.2 Calculate P_u and e in Fig. 7.23 and c of Problem 7.1 if the stresses in the tension steel are zero.

7.3 Determine P_{ub} and e_b for the rectangular cross sections of Fig. 7.23 in Problem 7.1.

7.4 For the cross sections shown in Fig. 7.23 of Problem 7.1, determine the load carrying capacities of the column sections for the following eccentricities to the plastic centroid
  (a) $e = 4$ in. (102 mm)
  (b) $e = 16$ in. (406 mm)

CHAPTER

8

Footings

8.1 INTRODUCTION

Cumulative floor loads of a superstructure are supported by foundation substructures in direct contact with the soil. The function of the foundation is to transmit safely the high concentrated column and/or wall reactions or lateral loads from earth-retaining walls to the ground without causing unsafe differential settlement of the supported structural system or soil failure.

If the supporting foundations are not adequately proportioned, one part of a structure can settle more than an adjacent part. Various members of such a system become overstressed at the column–beam joints due to *uneven* settlement of the supports, leading to large deformations. The additional bending and torsional moments in excess of the resisting capacity of the members can lead to excessive cracking due to yielding of the reinforcement and ultimately to failure.

If the total structure undergoes *even* settlement, little or no overstress occurs. Such behavior is observed when the foundation is excessively rigid and the supporting soil highly yielding such that a structure behaves similar to a floating body that can sink or tilt without breakage. Numerous examples of such structures can be found in such locations as Mexico City with buildings on mat foundations or rigid supports that sank several feet over the years due to the high consolidation of the supporting soil. Examples of other famous cases of the very slow and relatively uneven consolidation process can be cited. Gradual loss of stability of a structure undergoing tilting with time, such as the leaning Tower of Pisa, is an example of foundation problems resulting from uneven bearing support.

Layouts of structural supports vary widely and soil conditions differ from site to site and within a site. As a result, the type of foundation to be selected should be governed by these factors and by optimal cost considerations. In summary, the structural engineer needs to acquire the maximum economically feasible soil data on the site before embarking on a study of various possible alternatives for site layout.

A basic knowledge of soil mechanics and foundation engineering is assumed in presenting the topic of design of footings in this chapter. Background knowledge of the methodology of determining the resistance of cohesive and noncohesive soils is necessary to select the appropriate bearing capacity value for the particular site and the particular foundation system under consideration.

The bearing capacity of soils is usually determined by borings, test pits, or other soil investigations. If these are not available for the preliminary design, representative values at the footing level from Table 8.1 can normally be used.

TABLE 8.1   PRESUMPTIVE BEARING CAPACITY (TONS/FT^2)

Type of soil	Bearing capacity
Massive crystalline bedrock, such as granite, diorite, gneiss, and trap rock	100
Foliated rocks, such as schist or slate	40
Sedimentary rocks, such as hard shales, sandstones, limestones, and siltstones	15
Gravel and gravel–sand mixtures (GW and GP soils)	
Densely compacted	5
Medium compacted	4
Loose, not compacted	3
Sands and gravely sands, well graded (SW soil)	
Densely compacted	$3\frac{3}{4}$
Medium compacted	3
Loose, not compacted	$2\frac{1}{4}$
Sands and gravely sands, poorly graded (SP soil)	
Densely compacted	3
Medium compacted	$2\frac{1}{2}$
Loose, not compacted	$1\frac{3}{4}$
Silty gravels and gravel–sand–silt mixtures (GM soil)	
Densely compacted	$2\frac{1}{2}$
Medium compacted	2
Loose, not compacted	$1\frac{1}{2}$
Silty sand and silt–sand mixtures (SM soil)	2
Clayey gravels, gravel–sand–clay mixtures, clayey sands, sand–clay mixtures (GC and SC soils)	2
Inorganic silts, and fine sands; silty or clayey fine sands and clayey silts, with slight plasticity; inorganic clays of low to medium plasticity; gravely clays; sandy clays; silty clays; lean clays (ML and CL soils)	1
Inorganic clays of high plasticity, fat clays; micaceous or diatomaceous fine sandy or silty soils, elastic silts (CH and MH soils)	1

8.2 *TYPES OF FOUNDATIONS*

Basically there are six types of foundation substructures, as shown in Fig. 8.1. The foundation area must be adequate to carry the column loads, the footing weight, and any overburden weight within the permissible soil pressure.

1. *Wall footings.*   Such footings constitute a continuous slab strip along the length of the wall, having a width larger than the wall thickness. The projection of the slab footing is treated as a cantilever loaded up by the distributed soil pressure. The length of the projection is determined by the soil bearing pressure, with the critical section for bending being at the face of the wall. The main reinforcement is placed perpendicular to the wall direction.

2. *Independent isolated column footings.*   They consist of rectangular or square slabs of either constant thickness or sloping toward the cantilever tip. They are reinforced in both directions and are economical for relatively small loads or for footings on rock.

3. *Combined footings.*   Such footings support two or more column loads. They are necessary when a wall column has to be placed on a building line and the footing slab cannot project outside the building line. In such a case, an independent footing would be eccentrically loaded, causing apparent tension on the foundation soil.

    In order to achieve a relatively uniform stress distribution, the footing for the exterior wall column can be combined with the footing of the adjoining interior column. Additionally, combined footings are also used when the distance between adjoining columns is relatively small, such as in the case of corridor columns, when it becomes more economical to build a combined footing for the closely spaced columns.

4. *Cantilever or strap footings.*   These are similar to the combined footings except that the footings for the exterior and interior columns are built independently. They are joined by a strap beam to transmit the effect of the bending moment produced by the eccentric wall column load to the interior column footing area.

5. *Pile foundations.*   This type of foundation is essential when the supporting ground consists of structurally unsound layers of material to large depths. The piles may be driven either to solid bearing on rocks or hardpan or deep enough into the soil to develop the allowable capacity of the pile through skin frictional resistance or a combination of both. The piles could be either precast—and hence driven into the soil—or cast in place by drilling a caisson and subsequently filling it with concrete. The precast piles could be reinforced or prestressed concrete. Other types of piles are made of steel or treated wood. In all types the piles have to be provided with appropriately designed concrete caps reinforced in both directions.

6. *Raft or floating foundations.*   Such foundation systems are necessary when the allowable bearing capacity of soil is very low to great depths, making pile

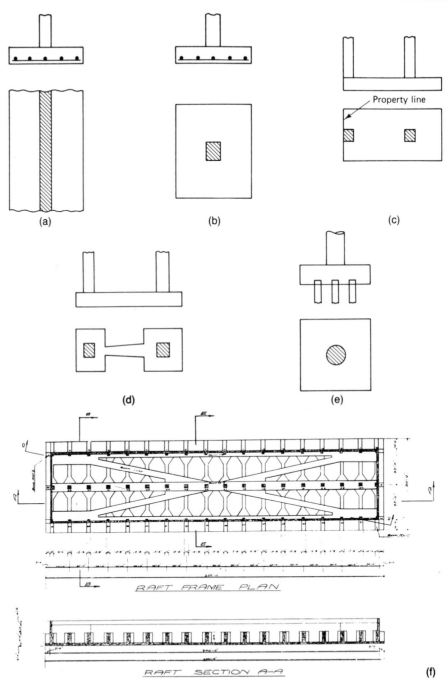

Figure 8.1 Types of foundations: (a) wall footing; (b) isolated footing; (c) combined footing; (d) strap footing; (e) pile foundation; (f) raft foundation.

foundations uneconomical. In this case, it becomes necessary to have a deep enough excavation with sufficient depth of soil removed that the net bearing pressure of the soil on the foundation is almost equivalent to the structure load. It becomes necessary to spread the foundation substructure over the entire area of the building such that the superstructure is considered to be theoretically floating on a raft. Continuously consolidating soils require such a substructure, which is basically an inverted floor system. Otherwise friction piles or piles driven to rock become mandatory.

8.3 SHEAR AND FLEXURAL BEHAVIOR OF FOOTINGS

To simplify foundation design, footings are assumed to be rigid and the supporting soil layers elastic. Consequently, uniform or uniformly varying soil distribution can be assumed. The net soil pressure is used in the calculation of bending moments and shears by subtracting the footing weight intensity and the surcharge from the total soil pressure. If a column footing is regarded as an *inverted* floor segment where the intensity of net soil pressure is considered to be acting as a column-supported cantilever slab, the slab would be subjected to both bending and shear in a similar manner to a floor slab subjected to gravity loads.

When heavy concentrated loads are involved, it has been found that shear rather than flexure controls most foundation designs. The mechanism of shear failure in footing slabs is similar to that in supported floor slabs. However, the shear capacity is considerably higher than that of beams. Because the footing in most cases bends in double curvature, shear and bending about both principal axes of the footing plan should be considered.

The state of stress at any element in the footing is due primarily to the combined effects of shear, flexure, and axial compression. Consequently, a basic understanding of the fundamental behavior of the footing slab and the cracking mechanism involved

Photo 56 Reinforced concrete footing after excavation. (Tests by F. E. Richart.)

is essential. It enables developing a background feeling for the underlying hypothesis used in the analysis and design requirements of footings both in shear and in flexure.

8.4 SOIL BEARING PRESSURE AT BASE OF FOOTINGS

The distribution of soil bearing pressure on the footing depends on the manner in which the column or wall loads are transmitted to the footing slab and the degree of rigidity of the footing. The soil under the footing is assumed to be a homogeneous elastic material and the footing is assumed to be rigid as a most common type of foundation. Consequently, the soil bearing pressure can be considered uniformly distributed if the reaction load acts through the axis of the footing slab area. If the load is not axial or symmetrically applied, the soil pressure distribution becomes trapezoidal due to the combined effects of axial load and bending.

8.4.1 Eccentric Load Effect on Footings

As indicated in Section 8.2, exterior column footings and combined footings can be subjected to eccentric loading. When the eccentric moment is very large, tensile stress on one side of the footing can result because the bending stress distribution depends on the magnitude of load eccentricity. It is always advisable to proportion the area of these footings so that the load falls within the middle kern, as shown in Figs. 8.2 and 8.3. In such a case, the location of the load is in the middle third of the footing dimension in each direction, thereby avoiding tension in the soil that can theoretically occur prior to stress redistribution.

1. *Eccentricity case $e < L/6$* (Fig. 8.2a). In this case, the direct P/A_f is larger than the bending stress M_c/I. The stress

$$p_{max} = \frac{P}{A_f} + \frac{Pe_1c}{I}$$ (8.1a)

$$p_{max} = \frac{P}{A_f} + \frac{Pe_1c}{I}$$ (8.1b)

2. *Eccentricity case $e_2 = L/6$* (Fig. 8.2b):

$$\text{Direct stress} = \frac{P}{A_f} = \frac{P}{sL}$$ (8.2a)

$$\text{Bending stress} = \frac{M_c}{I} = Pe_2 \times \frac{c}{I}$$ (8.2b)

$$\frac{c}{I} = \frac{L/2}{s(L^3/12)} = \frac{1}{s(L^2/6)} = \frac{6}{sL^2}$$ (8.2c)

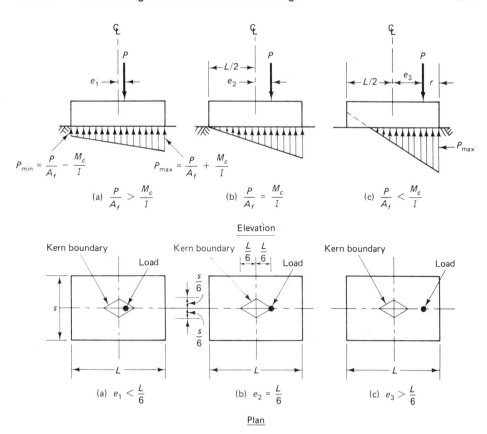

Figure 8.2   Eccentrically loaded footings.

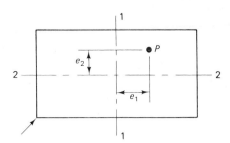

Figure 8.3   Biaxial loading of footing.

Here s and L are the width and the length of the footing, respectively. In order to find the limiting case where *no* tension exists on the footing, the direct stress P/A_f needs to be equivalent to the bending stress so that

$$\frac{P}{A_f} - P e_2 \frac{c}{I} = 0 \tag{8.2d}$$

Substituting for P/A_f and C/I from Eqs. 8.2a and 8.2c into Eq. 8.2d,

$$\frac{P}{sL} - Pe_2 \times \frac{6}{sL^2} = 0 \quad \text{or} \quad e_2 = \frac{L}{6}$$

Consequently, the eccentric load must act within the middle third of the footing dimension to avoid tension on the soil.

3. *Eccentricity case $e_3 > L/6$* (Fig. 8.2c). Because the load acts outside the middle third, tensile stress results at the left side of the footing, as shown in Fig. 8.2c. If the maximum bearing pressure p_{max} due to load P does not exceed the allowable bearing capacity of the soil, no uplift is expected at the left end of the footing and the center of gravity of the triangular bearing stress distribution *coincides* with the point of action of load p in Fig. 8.2c.

   The distance from the load P to the tip of footing is $r = (L/2) - e_3 = $ distance of the centroid of the stress triangle from the base of the triangle. Therefore the width of the triangle is $3r = 3[(L/2) - e_3]$. So the maximum compressive bearing stress is

$$p_{max} = \frac{P}{(3r \times s)/2} = \frac{2P}{3s[(L/2) - e_3]} \tag{8.3a}$$

4. *Eccentricity about two axes, biaxial loading* (Fig. 8.3). When a concentrated load has an eccentricity in two directions (both within their respective kern points), the stresses are

$$p_{max} = \frac{P}{A_f} \pm \frac{Pe_1c_1}{I_1} \pm \frac{Pe_2c_2}{I_2} \tag{8.3b}$$

8.4.2 Example 8.1: Concentrically Loaded Footings

A column support transmits axially a total service load of 400,000 lb (1779 kN) to a square footing at the frost line (3 ft below grade), as shown in Fig. 8.4. The frost line is the subgrade soil level below which the groundwater does not freeze throughout the year. Test borings indicate a densely compacted gravel–sand soil. Determine the required area of the footing and the net soil pressure intensity p_n to which it is subjected. Given:

$$\text{Unit weight of soil } \gamma = 135 \text{ lb/ft}^3 \text{ (21.1 kN/m}^3)$$

$$\text{Footing slab thickness} = 2 \text{ ft (0.61 m)}$$

Solution

Because the footing is concentrically loaded, the soil bearing pressure is considered uniformly distributed, assuming that the footing is rigid. From the soil test borings and Table 8.1, the presumptive bearing capacity of the soil is 5 tons/ft² at the level of the footing—that is, 10,000 lb/ft² (478.8 kPa). Assume that the average weight of the

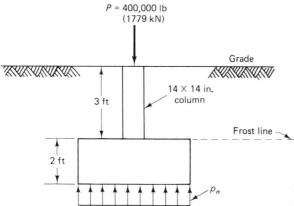

Figure 8.4   Concentrically loaded footing.

soil and concrete above the footing is 135 pcf. Because the top of the footing must be below the frost line (minimum 3 ft below grade), the net allowable pressure

$$p_n = 10,000 - (5 \times 135 + 100 \text{ psf for surcharge paving}) = 9225 \text{ psf}$$

$$\text{Minimum area of footing } A_f = \frac{400,000}{9,225} = 43.36 \text{ ft}^2$$

Use square footing 6 ft 8 in. × 6 ft 8 in. (2.03 m × 2.03 m),

$$A_f = 44.44 \text{ ft}^2 (4.13 \text{ m}^2) > 43.36 \text{ ft}^2$$

8.4.3 Example 8.2: Eccentrically Loaded Footings

A reinforced concrete footing supports a 14 in. × 14 in. column reaction $P = 400,000$ lb (1779 kN) at the frost line (3 ft below grade). The load acts at an eccentricity $e_1 = 0.4$ ft, $e_2 = 1.3$ ft, and $e_3 = 2.2$ ft. Select the necessary area of footing, assuming that it is rigid and has a thickness $h = 2\frac{1}{2}$ ft. Soil test borings have indicated that the bearing area is composed of layers of shale and clay to a considerable depth below the foundation. Use a unit weight $\gamma = 140$ lb/ft^3.

Solution

Using Table 8.1, assume an allowable bearing capacity $p_g = 6.5$ tons/ft^2 (13,000 lb/ft^2) at the footing base level.

Eccentricity $e_1 = 0.4$ ft

By trial and adjustment, assume a footing 5 ft × 9 ft (1.52 m × 2.74 m), $A_f = 45$ ft^2. Assume that the footing base is 6 ft below grade and that a slab on grade surcharge weighs 120 psf. Assume that the average weight of the soil and footing is ≈ 140 pcf.

$$\text{Net allowable bearing pressure } p_n = 13,000 - (6 \times 140 + 120)$$

$$= 12,040 \text{ lb/ft}^2 (576.5 \text{ KPa})$$

Stress due to the service eccentric column load is

$$p = \frac{P}{A_f} \pm \frac{Pe}{I/c} = \frac{400{,}000}{45} \pm \frac{400{,}000 \times 0.4 \times 6}{5(9)^2}$$

$$= 8889 \pm 2370 = 11{,}259 \text{ lb/ft}^2 \text{ (C) and } 6{,}519 \text{ lb/ft}^2 \text{ (C)} < 12{,}040 \text{ lb/ft}^2$$

The distribution of the bearing pressure is as shown in Fig. 8.5a—therefore O.K.

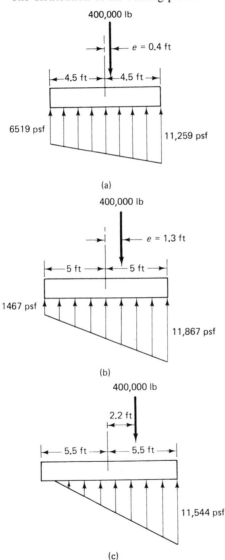

Figure 8.5 Bearing area and bearing stress distribution in Ex. 8.2.

Eccentricity $e_2 = 1.3$ ft

By trial and adjustment, assume a footing 6 ft × 10 ft (1.83 m × 3.05 m), $A_f = 60$ ft² (5.57 m²). The actual service load-bearing pressure is

$$p = \frac{400,000}{60} \pm \frac{400,000 \times 1.3 \times 6}{6(10)^2}$$

$$= 6667 \pm 5200 = 11,867 \text{ lb/ft}^2 \text{ (C) and } 1467 \text{ lb/ft}^2 \text{ (C)}$$

$$< 12,040 \text{ lb/ft}^2 \qquad \text{Therefore O.K.}$$

Notice in comparing the two cases that as the moment increases, leading to larger eccentricities, the minimum bearing pressure decreases, as seen from Fig. 8.5a and b.

Eccentricity $e_3 = 2.2$ ft

Using trial and adjustment, try a footing 7 ft × 11 ft (2.13 m × 3.35 m), $A_f = 77$ ft^2 (7.15 m^2).

$$p = \frac{400,000}{77} \pm \frac{400,000 \times 2.2 \times 6}{7(11)^2}$$

$$= 5195 \pm 6234 = 11,429 \text{ lb/ft}^2 \text{ (C) and } -1039 \text{ lb/ft}^2 \text{ (T)}$$

Check by Eq. 8.3 for $e > L/6 > 11.0/6 = 1.83$ ft:

$$p = \frac{2P}{3_s[(L/2) - e_3]} = \frac{2 \times 400,000}{3 \times 7[(11/2) - 2.2]}$$

$$= 11,544 \text{ lb/ft}^2 < 12,040 \text{ lb/ft}^2 \qquad \text{O.K.}$$

Figure 8.5c shows that because the load acts outside the middle third of the base, only part of the footing is subjected to compressive bearing stress.

8.5 DESIGN CONSIDERATIONS IN FLEXURE

The maximum external moment on any section of a footing is determined on the basis of computing the factored moment of the forces acting on the entire area of footing on *one side* of a vertical plane assumed to pass through the footing. This plane is taken at the following locations:

1. At the face of column, pedestal, or wall for an isolated footing as in Fig. 8.6a
2. Halfway between the middle and edge of wall for footing supporting masonry wall as in Fig. 8.6b
3. Halfway between face of column and edge of steel base for footings supporting a column with steel base plates

8.5.1 Reinforcement Distribution

The flexural reinforcement should be uniformly distributed across the entire width of the footing in one-way footings and two-way square footings. This recommendation is conservative, particularly if the soil bearing pressure is not uniform. No meaningful

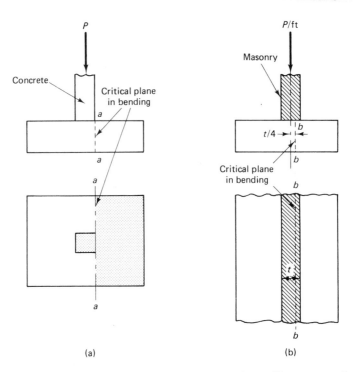

Figure 8.6 Critical planes in flexure: (a) concrete column; (b) masonry wall.

saving can be accomplished, however, if refinement is made in the bending moment assumptions.

In two-way rectangular footings supporting one column, the bending moment in the short direction is taken as equivalent to the bending moment in the long direction. The distribution of reinforcement differs in the long and short directions. The effective depth is assumed without meaningful loss of accuracy to be equal in both the short and long directions, although it differs slightly because of the two-layer reinforcing mats. Here is the recommended reinforcement distribution:

1. Reinforcement in the long direction is to be uniformly distributed across the entire width of the footing.
2. For reinforcement in the short direction, a central band of width equal to the width of footing in the short direction shall contain a major portion of the reinforcement total area in Eq. 8.4 uniformly distributed along the bandwidth:

$$\frac{\text{Reinforcement in bandwidth}}{\text{Total reinforcement in short direction, } A_s} = \frac{2}{\beta + 1} \qquad (8.4)$$

where β is the ratio of long to short side of footing. The remainder of the reinforcement required in the short direction is uniformly distributed outside the center band of the footing.

In all cases, the depth of the footing above the reinforcement must be at least 6 in. (152 mm) for footings on soil and at least 12 in. (305 mm) for footings on piles (footings on piles must always be reinforced). A practical depth for column footings should not be less than 9 in. (229 mm).

8.6 DESIGN CONSIDERATIONS IN SHEAR

The behavior of footings in shear does not differ from that of beams and supported slabs. Consequently, the same principles and expressions as those used in Chapter 5 on shear and diagonal tension are applicable to the shear design foundations.

The shear strength of slabs and footings in the vicinity of column reactions is governed by the more severe of the following two conditions.

8.6.1 Beam Action

The critical section for shear in slabs and footings is assumed to extend in a plane across the entire width and located at a distance d from the face of the concentrated load or reaction area. In this case, if only shear and flexure act, the nominal shear strength of the section is

$$V_c = 2\sqrt{f'_c}\, b_w d \tag{8.5}$$

where b_w is the footing width.

V_c must always be larger than the nominal shear force $V_n = V_u/\phi$ unless shear reinforcement is provided.

8.6.2 Two-Way Action

The plane of the critical section perpendicular to the plane of the slab is assumed to be so located that it has a minimum perimeter b_o. This critical section need not be closer than $d/2$ to the perimeter of the concentrated load or reaction area. The fundamental shear failure mechanism in two-way action demonstrates that the critical section occurs at a distance $d/2$ from the face of the support and not at d as in beam action.

The shear strength of the section in this case is

$$V_c = \left(2 + \frac{4}{\beta_c}\right)\sqrt{f'_c}\, b_o d \le 4\sqrt{f'_c}\, b_o d \tag{8.6}$$

where $\beta_c = \dfrac{\text{long side } c_l}{\text{short side } c_s}$ of the concentrated load or reaction area

b_o = perimeter of the critical section—that is, the length of the idealized failure plane

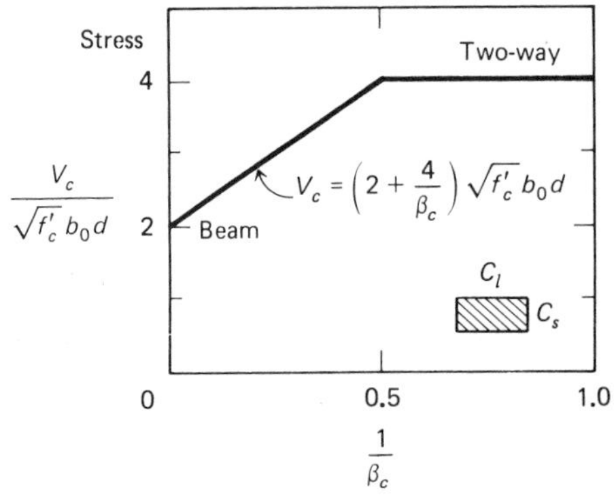

Figure 8.7   Shear strength in footings.

Figure 8.7 gives the relationship of the column side ratio β_c to the shear strength V_c of the footing. V_c must always be larger than the nominal shear force $V_n = V_u/\phi$ unless shear reinforcement is provided.

In cases of both one-way and two-way action, if shear reinforcement consisting of bars or wires is used, then

$$V_n = V_c + V_s \le 6\sqrt{f'_c}\,b_o d \qquad (8.7)$$

where $V_c = 2\sqrt{f'_c}\,b_o d$ and V_s is based on the shear reinforcement size and spacing as described in Chapter 5 unless shear heads made from steel I or channel shapes are used.

It is worthwhile to keep in mind that in most footing slabs, as in most supported superstructure slabs or plates, the use of shear reinforcement is not popular because of practical reasons and the difficulty of holding the shear reinforcement in position.

8.6.3 Force and Moment Transfer at Column Base

The forces and moments at the base of a column or wall are transferred to the footing by bearing on the concrete and by reinforcement, dowels, and mechanical connectors. Such reinforcement can transmit the compressive forces that exceed the concrete bearing strength of the footing or the supported column as well as any tensile force across the interface.

The permissible bearing stress on the actual loaded area of the column base or footing top area of contact is

$$f_b = \phi(0.85f'_c) \qquad \text{where } \phi = 0.70 \qquad (8.8a)$$

or $$f_b = 0.60f'_c \qquad (8.8b)$$

So the permissible bearing stress on the column can normally be considered $0.60f'_c$ for the column concrete. The compressive force that *exceeds* that developed by the permissible bearing stress at the base of the column or at the top of the footing must be carried by dowels or extended longitudinal bars.

If the footing supporting surface is wider on all sides than the loaded area, the code allows the design bearing strength on the loaded area to be multiplied by $\sqrt{A_2/A_1}$, but the value of $\sqrt{A_2/A_1}$ cannot exceed 2. A_1 is the loaded area and A_2 is the maximum area of the supporting surface that is geometrically similar and concentric with the loaded area.

A minimum area of reinforcement of $0.005A_g$ (but not less than four bars) has to be provided across the interface of the column and the footing even when the concrete bearing strength is not exceeded, A_g (in.²) being the gross area of the column cross section.

Lateral forces due to horizontal normal loads, wind, or earthquake can be resisted by shear-friction reinforcement.

8.7 OPERATIONAL PROCEDURE FOR THE DESIGN OF FOOTINGS

The following sequence of steps can be used for the selection and geometrical proportioning of the size and reinforcement spacing in footings.

1. Determine the allowable bearing capacity of the soil based on site boring test data and soil investigations.
2. Determine the service loads and bending moments acting at the base of the columns supporting the superstructure. Select the controlling service load and moment combinations.
3. Calculate the required area of the footing by dividing the total controlling service load by the selected allowable bearing capacity of the soil if the load is concentric or by also taking into account the controlling bending stress if combined load and bending moments exist.
4. Calculate the factored loads and moments for the controlling loading condition and find the required nominal resisting values by dividing the factored loads and moments by the applicable strength reduction factors ϕ.
5. By trial and adjustment, determine the required effective depth d of the section that has adequate punching shear capacity at a distance d from the support face for one-way action and at a distance $d/2$ for two-way action such that

$$V_c = 2\sqrt{f_c'}\, b_w d$$

for one-way action and

$$V_c = \left(2 + \frac{4}{\beta_c}\right)\sqrt{f_c'}\, b_o d \leq 4\sqrt{f_c'}\, b_o d$$

for two-way action, where b_w is the footing width for one-way action and b_o is the perimeter of the failure planes in two-way action. Use an average value of d, for there are two reinforcing mats in the footing. If the footing is rectangular, check the beam shear capacity in each direction on planes at a distance d from the face of the column support.

6. Calculate the factored moment of resistance M_u on a plane at the face of the column support due to the controlling factored loads from that plane to the extremity of the footing. Find $M_n = M_u/(\phi = 0.9)$. Select a total reinforcement area A_s based on M_n and the applicable effective depth.

7. Determine the size and spacing of the flexural reinforcement in the long and short directions:

   (a) Distribute the steel uniformly across the width of the footing in the long direction.

   (b) Determine the portion A_{s1} of the total steel area A_s determined in step 6 for the short direction to be uniformly distributed over the central band:

$$A_{s1} = \frac{2}{\beta + 1} A_s$$

   Distribute uniformly the remainder of the reinforcement $(A_s - A_{s1})$ outside the center band of the footing. Verify that the area of steel in each principal direction of the footing plan exceeds the minimum value required for temperature and shrinkage: $A_s = 0.0018b_w d$ for sections reinforced with grade 60 steel and $0.0020b_w d$ with grade 40 steel.

8. Check the development length and anchorage available to verify that bond requirements are satisfied (see Chapter 4).

9. Check the bearing stresses on the column and the footing at their area of contact such that the bearing strength P_{nb} for both is larger than the nominal value of column reaction $P_n = P_u/(\phi = 0.70)$. For footing bearing,

$$P_{nb} = \sqrt{A_2/A_1}\,(0.85f'_c A_1),$$

$\sqrt{A_2/A_1}$ not to exceed 2.

10. Determine the number and size of the dowel bars that transfer the column load to the footing slab.

Figure 8.8 presents a flowchart for the sequence of calculation operations.

8.8 EXAMPLES OF FOOTING DESIGN

8.8.1 Example 8.3: Design of Two-Way Isolated Footing

Design the footing thickness and reinforcement distribution for the isolated square footing in Ex. 8.1 if the total service load $P = 400,000$ consists of 230,000 lb (1023 kN) dead load and 170,000 lb (756 kN) live load. Given:

$f'_c = 3000$ psi (20.68 MPa), normalweight concrete (footing)

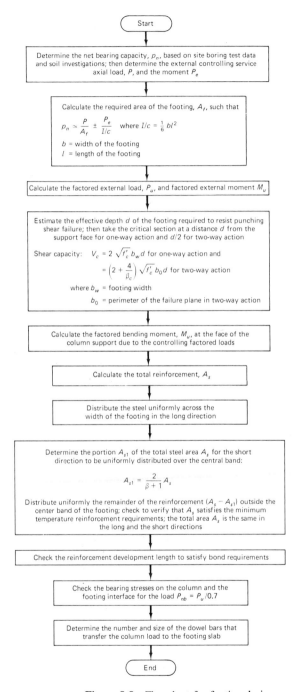

Figure 8.8 Flowchart for footing design.

Start

Determine the net bearing capacity, p_n, based on site boring test data and soil investigations; then determine the external controlling service axial load, P, and the moment P_e

Calculate the required area of the footing, A_f, such that

$$p_n \simeq \frac{P}{A_f} \pm \frac{P_e}{I/c} \quad \text{where } I/c = \frac{1}{6} bl^2$$

b = width of the footing
l = length of the footing

Calculate the factored external load, P_u, and factored external moment M_u

Estimate the effective depth d of the footing required to resist punching shear failure; then take the critical section at a distance d from the support face for one-way action and $d/2$ for two-way action

Shear capacity: $V_c = 2\sqrt{f'_c}\, b_w d$ for one-way action and

$$= \left(2 + \frac{4}{\beta_c}\right)\sqrt{f'_c}\, b_0 d \text{ for two-way action}$$

where b_w = footing width

b_0 = perimeter of the failure plane in two-way action

Calculate the factored bending moment, M_u, at the face of the column support due to the controlling factored loads

Calculate the total reinforcement, A_s

Distribute the steel uniformly across the width of the footing in the long direction

Determine the portion A_{s1} of the total steel area A_s for the short direction to be uniformly distributed over the central band:

$$A_{s1} = \frac{2}{\beta + 1} A_s$$

Distribute uniformly the remainder of the reinforcement $(A_s - A_{s1})$ outside the center band of the footing; check to verify that A_s satisfies the minimum temperature reinforcement requirements; the total area A_s is the same in the long and the short directions

Check the reinforcement development length to satisfy bond requirements

Check the bearing stresses on the column and the footing interface for the load $P_{nb} = P_u/0.7$

Determine the number and size of the dowel bars that transfer the column load to the footing slab

End

Figure 8.8 Flowchart for footing design.

$$f'_c = 5,500 \text{ psi } (37.91 \text{ MPa}) \text{ in column}$$

$$f_y = 60,000 \text{ psi } (413.7 \text{ MPa})$$

Solution

Factored load intensity (Step 4)

Data from Ex. 8.1:

Column size = 14 in. × 14 in. (355.6 mm × 355.6 mm)

Footing area = 6 ft 8 in. × 6 ft 8 in. (2.03 m × 2.03 m), $A_f = 44.49 \text{ ft}^2$

Assumed footing slab thickness $h = 2$ ft

Factored load $U = 1.4 \times 230,000 + 1.7 \times 170,000 = 611,000$ lb

Factored load intensity $= q_s = \dfrac{U}{A_f} = \dfrac{611,000}{44.49} = 13,733 \text{ lb/ft}^2 \text{ (657.6 kPa)}$

Shear capacity (Step 5)

Assume that the thickness of the footing slab ≈ 2 ft. The average depth $d = h - 3$ in. minimum cover ≈ 0.75 ≈ 20 in.

Beam action (at d from support face): The area to be considered for factored shear V_u is shown as *ABCD* in Fig. 8.9.

$$\text{Factored } V_u = 13,733\left(\frac{6 \text{ ft 8 in.}}{2} - \frac{14}{2 \times 12} - \frac{20}{12}\right)(6 \text{ ft 8 in.}) = 99,340 \text{ lb}$$

$$\text{Required } V_n = \frac{V_u}{\phi} = \frac{99,340}{0.85} = 116,871 \text{ lb}$$

$$b_w = 6 \text{ ft 8 in.} = 80 \text{ in. } (7.82 \text{ m})$$

$$\text{Available } V_c = 2\sqrt{f'_c}\, b_w d = 2\sqrt{3000} \times 80 \times 20 = 175,271 \text{ lb}$$

Two-way action (at d/2 from support face): The area to be considered for factored shear V_u is equal to the total area of footing less area *EFGH* of the failure zone.

$$\text{Factored } V_u = 13,733\left[44.49 - \left(\frac{14 + 20}{12}\right)^2\right] = 500,736 \text{ lb}$$

$$\text{Required } V_n = \frac{V_u}{\phi} = 589,100 \text{ lb } (2620 \text{ kN})$$

$$b_o = \text{perimeter of failure zone } EFGH = (14 + 20)4 = 136 \text{ in.}$$

$$\beta_c = \frac{14}{14} = 1.0$$

The available nominal shear strength from Eq. 8.6:

$$V_c = \left(2 + \frac{4}{\beta_c}\right)\sqrt{f'_c}\, b_o d \le 4\sqrt{f'_c}\, b_o d$$

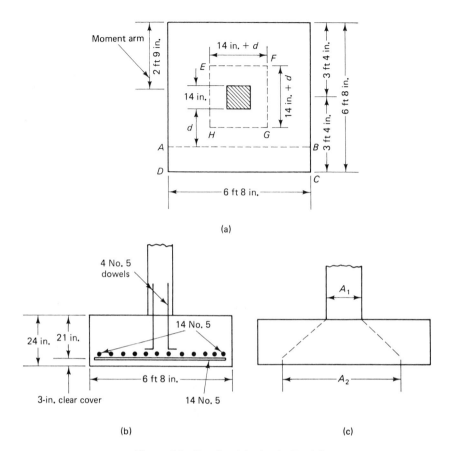

Figure 8.9   Details of footing in Ex. 8.3.

Because $\beta_c = 1.0$, $V_c = 4\sqrt{f'_c}\,b_o d$ has the lower value, hence controls.

$$V_c = 4\sqrt{3000} \times 136 \times 20 = 595{,}922 \text{ lb } (2650.7 \text{ kN}) > 589{,}100 \qquad \text{O.K.}$$

Therefore $d = 20$ in. is adequate for shear.

Bending moment capacity (Steps 6 and 7)

The critical section is at the face of the column.

$$\text{Moment arm} = \frac{6 \text{ ft } 8 \text{ in.}}{2} - \frac{14}{2 \times 12} = 2 \text{ ft } 9 \text{ in.}$$

$$\text{Factored moment } M_u = 13{,}733 \times 6.67 \left[\frac{(2 \text{ ft } 9 \text{ in.})^2}{2} \right]$$

$$= 346{,}359.1 \text{ ft-lb} = 4{,}156{,}310 \text{ in.-lb}$$

$$M_n = \frac{M_u}{\phi} = \frac{4,156,310}{0.90} = 4,618,122 \text{ in.-lb (521.8 kN-m)}$$

$$M_n = A_s f_y \left(d - \frac{a}{2} \right)$$

Assume that $(d - a/2) \approx 0.9d$. Use average $d = 20$ in.

$$4,618,122 = A_s \times 60,000 \times 0.9 \times 20$$

or

$$A_s = \frac{4,618,122}{60,000 \times 0.9 \times 20} = 4.28 \text{ in.}^2/\text{80-in. band}$$

$$a = \frac{A_s f_y}{0.85 f'_c b} = \frac{4.28 \times 60,000}{0.85 \times 3000 \times 80} = 1.26 \text{ in.}$$

$$4,618,122 = A_s \times 60,000 \left(20 - \frac{1.26}{2} \right)$$

$$A_s = 3.98 \text{ in.}^2 \qquad \rho = \frac{A_s}{bd} = \frac{3.98}{80 \times 20} = 0.0025$$

Minimum allowable temperature and shrinkage steel:

$$\rho_{min} = 0.0018 < \rho \qquad \text{O.K.}$$

Use 14 No. 5 bars ($A_s = 4.27$ in.^{2}) each way, spaced at $\approx 5\frac{1}{2}$ in. (139.7 mm) center to center.

Development of reinforcement (Step 8)

The critical section for development-length determination is the same as the critical section in flexure—namely, at the face of the column. From Eq. 4.4a for No. 5 bars,

$$l_d = \frac{0.04 A_b f_y}{\sqrt{f'_c}}$$

$$\leq 0.0004 d_b f_y$$

Note that a reduction multiplier $\lambda_d = 0.8$ cannot be used because the bar spacing is less than 6 in. Or

$$l_d = \frac{0.04 \times 0.305 \times 60,000}{\sqrt{3000}} = 13.36 \text{ in.}$$

$$l_d = 0.0004 \times 0.625 \times 60,000 = 15.0 \text{ in. controls}$$

The projection length of each bar beyond the column face is

$$\frac{1}{2}(6 \text{ ft } 8 \text{ in.} - 14 \text{ in.}) - 3\text{-in. cover} = 30 \text{ in.} > 15 \text{ in.} \qquad \text{O.K.}$$

Force transfer at interface of column and footing (Step 9)

Column $f'_c = 5500$ psi. Factored $P_u = 611,000$ lb.

(a) Bearing strength on column, using Eq. 8.8b:

$$\phi P_{nb} = 0.70 \times 0.85 f_c' A_1 = 0.60 f_c' A_1$$

or

$$\phi P_{nb} = 0.60 f_c' A_1 = 0.60 \times 5500 \times 14 \times 14$$

$$= 646,800 \text{ lb} > 611,000 \qquad \text{O.K.}$$

From step 9 of the design operational procedure on bearing strength on footing concrete:

$$\sqrt{\frac{A_2}{A_1}} = \sqrt{\frac{(6 \text{ ft } 8 \text{ in.}) \times (6 \text{ ft } 8 \text{ in.})}{(14 \times 14)/144}} = 5.714 > 2.0 \qquad \text{Use 2}$$

$$\phi P_{nb} = 2.0(0.60 f_c' A_1) = 2 \times 0.60 \times 3000 \times 14 \times 14 = 705,600 \text{ lb}$$

$$> 611,000 \qquad \text{O.K.}$$

Dowel bars between column and footing (Step 10)

Even though the bearing strength at the interface between the column and the footing slab is adequate to transfer the factored P_u, a minimum area of reinforcement is necessary across the interface. The minimum $A_s = 0.005(14 \times 14) = 0.98$ in.^{2} but not less than four bars. Use four No. 5 bars as dowels ($A_s = 1.22$ in.^{2}).

Development of dowel reinforcement in compression: From Eqs. 4.5a and 4.5b, for No. 5 bars,

$$l_{db} = \frac{0.02 d_b f_y}{\sqrt{f_c'}}$$

and $l_{db} \geq 0.0003 d_b f_y$, where d_b is the dowel bar diameter. Within column:

$$l_d = \frac{0.02 \times 0.625 \times 60,000}{\sqrt{5500}} = 10.11 \text{ in.}$$

$$0.0003 \times 0.625 \times 60,000 = 11.25 \text{ in.} \qquad \text{controls}$$

Within footing:

$$l_d = \frac{0.02 \times 0.625 \times 60,000}{\sqrt{3000}} = 13.69 \text{ in.}$$

Available length for development above the footing reinforcement, assuming column bars size to be the same as the dowel bars size:

$$l = 24 - 3 \text{ (cover)} - 2 \times 0.625 \text{ (footing bars)} - 0.625 \text{ (dowels)}$$

$$= 19.13 > 13.69 \text{ in.} \qquad \text{O.K.}$$

8.8.2 Example 8.4: Design of Two-Way Rectangular Isolated Footing

Determine the size and distribution of the bending reinforcement of an isolated rectangular footing subjected to a concentrated concentric factored column load

P_u = 770,000 lb (3425 kN) and having an area 10 ft × 15 ft (3.05 m × 4.57 m). Given:

$$f'_c = 3000 \text{ psi (20.68 MPa), footing}$$

$$f_y = 60,000 \text{ psi (413.7 MPa)}$$

$$\text{Column size} = 14 \text{ in.} \times 18 \text{ in.}$$

Solution

$$\text{Factored load intensity } q_s = \frac{770,000}{10 \times 15} = 5134 \text{ lb/ft}^2$$

Shear capacity (Step 5)

Through trial and adjustment, assume that the footing slab is 2 ft 4 in. thick.

Beam action (at distance d from column face): Average effective depth ≈ 2 ft 4 in. − 3 in. (cover) − $\frac{3}{4}$ in. (diameter of bars in first layer) ≈ 24 in.

From Fig. 8.10, length *CD* subjected to bearing intensity q_s in one-way beam action:

$$\frac{15 \text{ ft}}{2} - \frac{18 \text{ in.}}{2 \times 12} - \frac{24 \text{ in.}}{12} = 4 \text{ ft 9 in.} = 57 \text{ in.}$$

$$\text{Factored } V_u = 5134 \times 10 \text{ ft} \times 4 \text{ ft 9 in.} = 243,865 \text{ lb}$$

$$\text{Required } V_n = \frac{V_u}{\phi} = \frac{243,865}{0.85} = 286,900 \text{ lb}$$

$$\text{Available } V_n = 2\sqrt{f'_c}\,b_w d = 2\sqrt{3000} \times 120 \times 24$$

$$= 315,488 \text{ lb} > 286,900 \quad \text{O.K.}$$

Notice that the shorter side length was used for b_w to give the lower available V_n value.

Two-way action (at distance d/2 from column face):

Loaded area outside the failure zone *LMNP* in Fig. 8.10

$$= 15 \times 10 - (c_l + d)(c_s + d)$$

$$= 150 - \frac{(18 + 24)(14 + 24)}{144}$$

$$= 138.92 \text{ ft}^2$$

$$\text{Factored } V_u = 5134 \times 138.92 = 713,215 \text{ lb}$$

$$\text{Required } V_n = \frac{713,215}{0.85} = 839,077 \text{ lb (3732 kN)}$$

$$\text{Perimeter of shear failure plane } b_o = 2[(c_1 + d) + (c_s + d)]$$

$$= 2[(18 + 24) + (14 + 24)] = 160 \text{ in.}$$

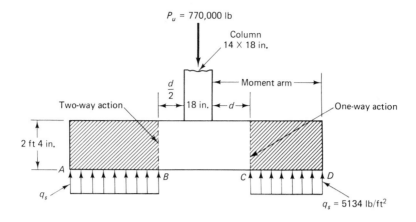

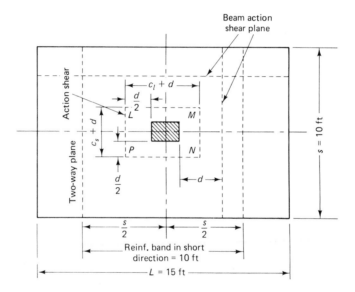

Figure 8.10 Beam action and two-way action planes in Ex. 8.4.

From Eq. 8.6

$$V_c = \left(2 + \frac{4}{\beta_c}\right)\sqrt{f_c'}\,b_o d \le 4\sqrt{f_c'}\,b_o d$$

$$\beta_c = \frac{18}{14} = 1.286 \qquad 2 + \frac{4}{1.286} > 4$$

Thus $V_c = 4\sqrt{f_c'}\,b_o d$ controls.

$$\text{Available } V_c = 4\sqrt{f_c'}\, b_o d = 4\sqrt{3000} \times 160 \times 24$$

$$= 841{,}302 \text{ lb} > 839{,}077 \qquad \text{O.K.}$$

Design of two-way reinforcement

The critical section for bending is at the face of the column. The controlling moment arm is in the long direction:

$$\frac{15 \text{ ft}}{2} - \frac{18 \text{ in.}}{2 \times 12} = 6.75 \text{ ft (2.06 m)}$$

$$\text{Factored moment } M_u = 5134 \times \frac{10(6.75)^2}{2}$$

$$= 1{,}169{,}589 \text{ ft-lb} = 14{,}035{,}073 \text{ in.-lb (1585.96 kN-m)}$$

$$M_n = \frac{14{,}035{,}073}{0.9} = 15{,}594{,}526 \text{ in.-lb (1762.18 kN-m)}$$

Assume that $(d - a/2) \simeq 0.9d$.

$$M_n = A_s f_y \left(d - \frac{a}{2} \right) \quad \text{or} \quad 15{,}594{,}526 = A_s \times 60{,}000 \times 0.9 \times 24$$

$$A_s = \frac{15{,}594{,}526}{60{,}000 \times 0.9 \times 24} = 12.03 \text{ in.}^2/10\text{-ft-wide strip}$$

Check:

$$a = \frac{A_s f_y}{0.85 f_c' b} = \frac{12.03 \times 60{,}000}{0.85 \times 3000 \times 120} = 2.36 \text{ in.}$$

$$15{,}594{,}526 = A_s \times 60{,}000 \left(24 - \frac{2.36}{2} \right)$$

$$A_s = 11.39 \text{ in.}^2 = \frac{11.39}{10 \text{ ft}} = 1.14 \text{ in.}^2/\text{ft width}$$

Try No. 8 bars, $A_s = 0.79 \text{ in.}^2$ per bar.

$$\text{Number of bars in the short direction} = \frac{11.39}{0.79} = 14.42$$

Use 15 bars.

Reinforcement in the short direction

The bandwidth $= s = 10$ ft (Fig. 8.10). From Eq. 8.4 we have

$$\beta = \frac{15}{10} = 1.5$$

$$\frac{A_{s1}}{A_s} = \frac{2}{\beta + 1} \quad \text{or} \quad \frac{A_{s1}}{11.39} = \frac{2}{1 + 1.5}$$

Therefore

$$A_{s1} = \frac{2 \times 11.39}{2.5} = 9.11 \text{ in.}^2$$

to be placed in the central 10-ft-wide band and the balance $(11.39 - 9.11 = 2.28 \text{ in.}^2)$ to be placed in the remainder of the footing. Use 12 No. 8 bars in the central band $= 9.48 \text{ in.}^2$ and two No. 8 bars at each side of the band, as in Fig. 8.11. To complete the design, a check of the development length, bearing stress at the column-footing interface, and dowel action has to be made, as in Ex. 8.3.

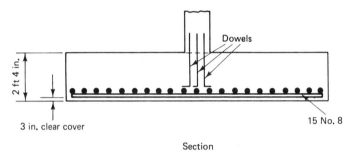

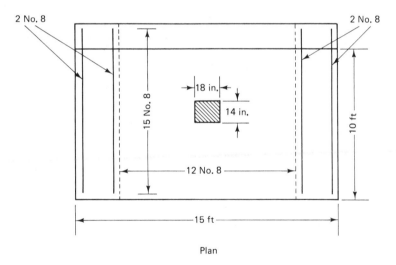

Figure 8.11   Footing reinforcement details of Ex. 8.4.

8.8.3 Example 8.5: Proportioning of a Combined Footing

A combined footing has the layout shown in Fig. 8.12. Column L at the property line is subjected to a total service axial load $P_L = 200,000$ lb (889.6 kN) and the internal column R is subjected to a total service load $P_R = 350,000$ lb (1556.8 kN). The live load is 35% of the total load. The bearing capacity of the soil at the level of the footing base is 4000 lb/ft^2 (191.5 kPa) and the average value of the soil and footing unit weight $\gamma = 120$ pcf (1922 kg/m^3). A surcharge of 100 lb/ft^2 results from the slab on grade. Proportion the footing size and select the necessary size and distribution of the footing slab reinforcement. Given:

$$f'_c = 3000 \text{ psi (20.68 MPa)}$$

$$f_y = 60,000 \text{ psi (413.7 MPa)}$$

Base of footing at 7 ft below grade

Solution

Total columns load $= 200,000 + 350,000 = 550,000$ lb (2446.4 kN)

Net allowable soil capacity $p_n = p_g - 120(7 \text{ ft height to base of footing}) - 100$

or

$$p_n = 4000 - 120 \times 7 - 100 = 3060 \text{ lb/ft}^2$$

$$\text{Minimum footing area } A_f = \frac{P}{p_n} = \frac{550,000}{3060} = 179.8 \text{ ft}^2$$

Center of gravity of column loads from the property line:

$$\bar{x} = \frac{200,000 \times 0.5 + 350,000 \times 20.5}{550,000} = 13.23 \text{ ft}$$

Length of footing $L = 2 \times 13.23 = 26.46$ ft

Use $L = 27$ ft.

$$\text{Width of footing } S = \frac{179.8}{27.0} = 6.66 \text{ ft}$$

Use $S = 6$ ft 6 in. as shown in Fig. 8.12.

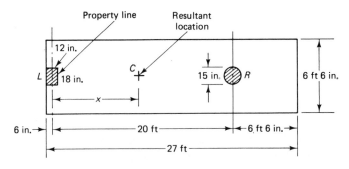

Figure 8.12 Combined footing plan geometry in Ex. 8.5.

Factored shears and moments

Column L:

$$P_D = 0.65 \times 200,000 = 130,000 \text{ lb}$$

$$P_L = 200,000 - 130,000 = 70,000 \text{ lb}$$

$$P_U = 1.4 \times 130,000 + 1.7 \times 70,000 = 301,000 \text{ lb}$$

Column R:

$$P_D = 227,500 \text{ lb}$$

$$P_L = 122,500 \text{ lb}$$

$$P_U = 1.4 \times 227,500 + 1.7 \times 122,500 = 526,750 \text{ lb}$$

The net factored soil bearing pressure for footing structural design is

$$q_s = \frac{P_u}{A_f} = \frac{301,000 + 526,750}{6.5 \times 27.0} = 4716.5 \text{ lb/ft}^2$$

Assume that the column loads are acting through their axes.

Factored bearing pressure per foot width $= q_s \times S = 4,716.5 \times 6.5 = 30,658 \text{ lb/ft}$

$$V_u \text{ at centerline of column } L = 301,000 - 30,658 \times \frac{6}{12} = 285,671 \text{ lb}$$

$$V_u \text{ at centerline of column } R = 526,750 - 30,658 \times 6.5 = 327,473 \text{ lb}$$

The maximum moment is at the point C of zero shear in Fig. 8.13 x (ft) from the center of the left column L.

$$x = \frac{285,671 \text{ lb}}{30,658 \text{ plf}} = 9.32 \text{ ft}$$

Taking a free-body diagram to the left of a section through C, the factored moment at point C is

$$M_{uc} = \frac{w_u l^2}{2} - P_{ul}x$$

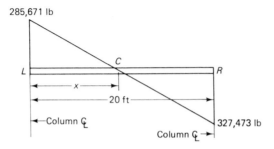

Figure 8.13   Shear diagram of footing in Ex. 8.5.

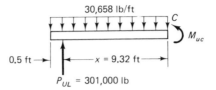

Figure 8.14   Free-body diagram.

$$M_u \text{ from left side} = 30,658 \frac{(9.32 + 0.50)^2}{2} - 301,000 \times 9.32$$

$$= -1,327,108 \text{ ft-lb} = -15,925,293 \text{ in.-lb} \qquad \text{(Fig. 8.14)}$$

$$M_u \text{ from right side} = 30,658 \frac{(27.0 - 9.82)^2}{2} - 526,750(20.0 - 9.32)$$

$$= 1,101,299 \text{ ft-lb} = -13,215,588 \text{ in.-lb}$$

Thus M_u from the left side controls. Note that M_u from the right side differs from M_u from the left side because the footing length of 27 ft is used instead of the computed length of 26.46 ft and because x is rounded off. Therefore the load is not exactly uniform due to the small eccentricity.

Design of the footing in the longitudinal direction

(a) *Shear:* The combined footing is considered as a beam in the shear computations. So the critical section is at a distance d from the face of the support. Controlling V_n at the column centerline

$$\frac{V_u}{\phi} = \frac{327,473}{0.85} = 385,262 \text{ lb}$$

Assume that the total footing thickness = 3 ft (0.92 m). The effective footing depth $d = 32$ in. for minimum steel cover ≈ 4 in. For the controlling interior column R, the equivalent rectangular column size $\sqrt{\pi(15)^2/4} = 13.29$ in.

$$\text{Required } V_n \text{ at } d \text{ section} = 385,262 - \frac{13.29/2 + d)}{12} \times \frac{30,658}{\phi}$$

$$= 385,262 - \frac{38.65 \times 30,658}{12 \times 0.85} = 269,107 \text{ lb (1196.9 kN)}$$

$$V_c = 2\sqrt{f_c'}\, b_w d = 2\sqrt{3000} \times 6.5 \times 12 \times 32$$

$$= 273,423 \text{ lb (1216.2 kN)} > 269,107 \qquad \text{O.K.}$$

(b) *Moment and reinforcement in the longitudinal direction (Step 4):* The distribution of shear and moment in the longitudinal direction is shown in Fig. 8.15. The critical section for moment is taken at the face of the columns.

$$\text{Controlling moment } M_n = \frac{M_u}{\phi} = \frac{15,925,293}{0.9} = 17,694,770 \text{ in.-lb (1999.5 kNm)}$$

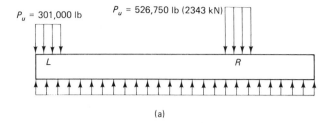

(a)

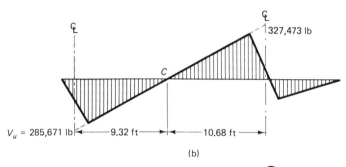

(b)

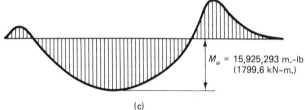

(c)

Figure 8.15   Longitudinal shear and moment distribution: (a) elevation; (b) shear; (c) moment.

$$M_n = A_s f_y \left(d - \frac{a}{2} \right)$$

Assume that $(d - a/2) \simeq 0.9d$.

$$17,694,770 = A_s \times 60,000(0.9 \times 32)$$

or

$$A_s = \frac{17,694,770}{60,000 \times 0.9 \times 32} = 10.24 \text{ in.}^2$$

$$a = \frac{A_s f_y}{0.85 f_c' b} = \frac{10.24 \times 60,000}{0.85 \times 3,000 \times 6.5 \times 12} = 3.09 \text{ in.}$$

$$17,694,770 = A_s \times 60,000 \left(32 - \frac{3.09}{2} \right)$$

$$A_s = 9.68 \text{ in.}^2 \ (6245 \text{ mm}^2)$$

Use 22 No. 6 bars at the top for the middle span.

$$A_s = 9.68 \text{ in.}^2 \ (22 \text{ bars } 19.1 \text{ mm diameter})$$

Design of footing in the transverse direction

Both columns are treated as isolated columns. The width of the band should not be larger than the width of the column plus half the effective depth d on *each* side of the column. This assumption is on the safe side, for the actual bending stress distribution is highly indeterminate. It is, however, possible to assume that the flexural reinforcement in the transverse direction can raise the shear punching capacity within the $d/2$ zone from the face of the rectangular left column L and the *equivalent* rectangular right column R. Figure 8.16 shows the transverse bandwidths for both columns L and R, determined on the basis of this discussion.

$$\text{Bandwidth } b_L = 12 + \frac{32}{2} = 28 \text{ in.} = 2.33 \text{ ft}$$

The rectangular column size equivalent to the circular interior 15-in.-diameter column = 13.29 in.

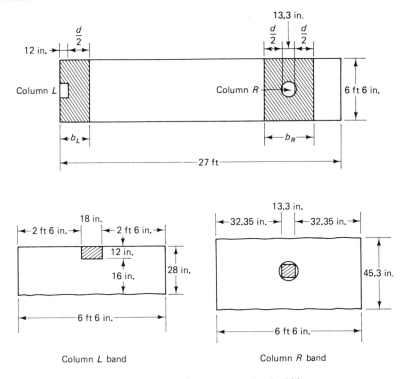

Figure 8.16   Footing transverse band widths.

$$\text{Bandwidth } b_R = 13.29 + 2\left(\frac{32}{2}\right) = 45.3 \text{ in. } = 3.77 \text{ ft}$$

Column L transverse band reinforcement:

$$\text{Moment arm} = \frac{6 \text{ ft } 6 \text{ in.}}{2} - \frac{18}{2 \times 12} = 2.50 \text{ ft} = 30.0 \text{ in.}$$

The net factored bearing pressure in the transverse direction,

$$q_s = \frac{301,000}{6.5} = 46,308 \text{ lb/ft}$$

$$M_u = q_s \frac{l^2}{2} = 46,308 \frac{(2.50)^2}{2} = 144,713 \text{ ft-lb} = 1,736,550 \text{ in.-lb}$$

$$M_n = \frac{M_u}{\phi} = \frac{1,736,550}{0.90} = 1,929,500 \text{ in.-lb } (218.0 \text{ kN-m})$$

$$M_n = A_s f_y \left(d - \frac{a}{2}\right)$$

or

$$1,929,500 = A_s \times 60,000 \times 0.9 \times 32$$

$$A_s = 1.12 \text{ in.}^2$$

$$a = \frac{A_s f_y}{0.85 f'_c b} = \frac{1.12 \times 60,000}{0.85 \times 3,000 \times 28} = 0.94 \text{ in.}$$

$$1,929,500 = A_s \times 60,000 \left(32 - \frac{0.94}{2}\right)$$

$$A_s = 1.02 \text{ in.}^2 \ (658 \text{ mm}^2)$$

$$\text{min. } A_s = 0.0018 b_w d = 0.0018 \times 28 \times 32 = 1.62 \text{ in.}^2$$

$$\rho = \frac{1.02}{28 \times 32} = 0.00114$$

Use six No. 5 bars, $A_s = 1.86$ in.^{2} (six bars 15.9 mm diameter) equally spaced in the band that is to be centered under the column.

Column R transverse band reinforcement: Equivalent square column size = 13.3 in. × 13.3 in.

$$\text{Moment arm} = \frac{6 \text{ ft } 6 \text{ in.}}{2} - \frac{13.3 \text{ in.}}{2 \times 12} = 2.69 \text{ ft} = 32.35 \text{ in.}$$

Net factored bearing pressure in the transverse direction,

$$q_s = \frac{526,750}{6.50} = 81,038 \text{ lb/ft}$$

$$M_u = q_s \frac{l^2}{2} = 81,038 \frac{(2.69)^2}{2} = 293,200 \text{ ft-lb} = 3,518,400 \text{ in.-lb}$$

$$M_n = \frac{M_u}{\phi} = \frac{3,518,400}{0.90} = 3,909,333 \text{ in.-lb (441.8 kN-m)}$$

$$M_n = A_s f_y \left(d - \frac{a}{2} \right) \qquad \text{Assume that } d - \frac{a}{2} \simeq 0.90d$$

or

$$3,909,333 = A_s \times 60,000 \times 0.9 \times 32$$

$$A_s = 2.26 \text{ in.}^2 \qquad a = \frac{A_s f_y}{0.85 f'_c b} = \frac{2.26 \times 60,000}{0.85 \times 3,000 \times 45.3} = 1.17 \text{ in.}$$

$$3,909,333 = A_s \times 60,000 \left(32 - \frac{1.17}{2} \right)$$

$$A_s = 2.07 \text{ in.}^2 \text{ (3347 mm}^2\text{)}$$

$$\rho = \frac{2.07}{45.3 \times 32} = 0.0014 < \rho_{min}$$

where $\rho_{min} = 0.0018$ (shrinkage temperature reinforcement).

$$\text{Minimum } A_s = 0.0018 \times 45.3 \times 32 = 2.61 \text{ in.}^2$$

Use nine No. 5 bars, $A_s = 2.79$ in.^{2} (nine bars 15.9 mm diameter) equally spaced.

Development length check for bars in tension

(a) *Longitudinal top steel:* From Eq. 4.4

$$\min l_{db} = \frac{0.04 A_b f_y (1.4)}{\sqrt{f'_c}}$$

or

$$l_{db} = 0.04 \times 0.44 \times 60,000 \times \frac{1.4}{\sqrt{3000}} = 26.99 \text{ in.} = 2.25 \text{ ft} \qquad \text{(governs)}$$

or

$$l_{db} = 0.0004 d_b f_y = 0.0004 \times 0.750 \times 60,000 = 18.0 \text{ in.} = 1.50 \text{ ft}$$

Modifying multiplier L_d for top reinforcement $= 1.4$; thus the minimum development length $l_{db} = 1.4 \times 26.99 = 37.79$ in. $= 3.15$ ft. The distance from point C at the maximum moment in Fig. 8.13 to the center of the left column $= 9.32 + 0.50 = 9.82$ ft > 3.15. O.K.

(b) *Transverse bottom steel:*

$$l_{db} = 0.04 \times 0.305 \times \frac{60,000}{\sqrt{3000}} = 13.36 \text{ in.}$$

$$l_{db} = 0.004 \times 0.625 \times 60,000 = 15.00 \text{ in.}$$

Available development length $= (32.25 - 3.0)$ in. > 15.00 in.   O.K.

$$\text{Required modifier for column } L = \frac{1.62}{1.86} = 0.87$$

$$\text{Required modifier for column } R = \frac{2.61}{2.79} = 0.94$$

Minimum development length $l_{db} = 0.94 \times 15.00 = 14.10$ in L available

Therefore adopt reinforcement as in Fig. 8.17. Check for dowel steel from the columns to the footing slab.

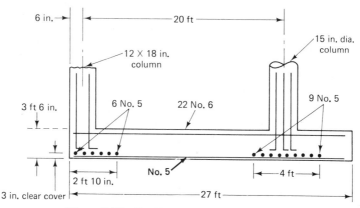

Figure 8.17 Combined footing reinforcement.

SELECTED REFERENCES

8.1 Furlong, R. W., "Design Aids for Square Footings," *Journal of the American Concrete Institute,* Proc. Vol. 62, March 1965, pp. 363–371.

8.2 Richart, F. E., "Reinforced Concrete Walls and Column Footings," *Journal of the American Concrete Institute,* Proc. Vol. 45, October and November 1948, pp. 97–127 and 237–245.

8.3 Sowers, G. B. and G. F. Sowers, *Introductory Soil Mechanics and Foundations,* 3rd ed., Macmillan, New York, 556 pp.

8.4 American Insurance Association, *The National Building Code,* 1976 edition, New York, December 1977, 767 pp.

PROBLEMS FOR SOLUTION

8.1 Design a reinforced concrete square isolated footing to support an axial column service live load P_L = 300,000 lb (1334 kN) and service dead load P_D = 625,000 lb (2780 kN). The size of the column is 30 in. × 24 in. (0.76 m × 0.61 m). The soil test borings indicate that it is composed of medium compacted sands and gravely sands, poorly graded. The frost line is assumed to be 3 ft below grade.

Given: Average weight of soil and concrete above the footing,

$$\gamma = 130 \text{ pcf } (20.41 \text{ kN/m}^3)$$
$$\text{Footing } f'_c = 3000 \text{ psi } (20.68 \text{ MPa})$$

$$\text{Column } f'_c = 4000 \text{ psi } (27.58 \text{ MPa})$$
$$f_y = 60,000 \text{ psi } (413.7 \text{ MPa})$$
$$\text{Surcharge} = 120 \text{ psf } (5.7 \text{ kPa})$$

8.2 Design a reinforced concrete wall footing for (a) 10-in. (0.25 m) reinforced concrete wall and (b) 12-in. (0.30 m) masonry wall. The intensity of service linear dead load is W_D = 20,000 lb/ft (292.0 kN/m) and a service linear live load W_L = 15,000 lb/ft (219.0 kN/m) of wall length. Assume an evenly distributed soil bearing pressure and that the average soil bearing pressure at the base of the footing is 3 tons/ft^2 (87.6 kN/m). The frost line is assumed to be 2 ft below grade.

Given: Average weight of soil and footing above base = 125 pcf (19.6 kN/m^3)

$$\text{Footing } f'_c = 3000 \text{ psi } \quad (20.68 \text{ MPa})$$
$$\text{Column } f'_c = 5000 \text{ psi } \quad (34.47 \text{ MPa})$$
$$f_y = 60,000 \text{ psi } (413.7 \text{ MPa})$$

8.3 A combined footing is subjected to an exterior 16-in. × 16-in. (0.4 m × 0.4 m) column abutting the property line, carrying a total service load P_W = 300,000 lb (1334.4 kN) and an interior column 20 in. × 20 in. (0.5 m × 0.5 m) carrying a total factored load P_w = 400,000 lb (1779.2 kN). The live load is 30% of the total load. The centerline distance between the two columns is 22 ft (6.71 m). Design the appropriate reinforced concrete footing on a soil weighing 135 pcf (21.2 kN/m^3). The bearing capacity of the soil at the level of the footing base is 6000 lb/sq ft. The frost line is assumed to be at 3 ft 6 in. (1.07 m) below grade. Assume a surcharge of 125 psf (19.62 kN/m^3) at grade level. Given:

$$\text{Footing } f'_c = 3500 \text{ psi } \quad (24.13 \text{ MPa})$$
$$\text{Column } f'_c = 5500 \text{ psi } \quad (37.42 \text{ MPa})$$
$$f_y = 60,000 \text{ psi } (413.7 \text{ MPa})$$

Geometry

To convert from	to	multiply by[†]
Length		
inch	millimeter (mm)	25.4E
foot	meter (m)	0.3048E
yard	meter (m)	0.9144E
mile(statute)	kilometer(km)	1.609
Area		
square inch	square centimeter (cm^2)	6.452
square foot	square meter (m^2)	0.09290
square yard	square meter (m^2)	0.8361
Volume (Capacity)		
ounce	cubic centimeter (cm^3)	29.57
gallon	cubic meter (m^3)[‡]	0.003785
cubic inch	cubic centimeter (cm^3)	16.4
cubic foot	cubic meter (m^3)	0.02832
cubic yard	cubic meter (m^3)[‡]	0.765
Force		
kilogram-force	newton (N)	9.807
kip-force	kilonewton (kN)	4.448
pound-force	newton (N)	4.448
Pressure or Stress (Force per Area)		
kilogram-force/square meter	pascal (Pa)	9.807
kip-force/square inch (ksi)	megapascal (MPa)	6.895
newton/square meter (N/m^2)	pascal (Pa)	1.000E
pound-force/square foot	pascal (Pa)	47.88
pound-force/square inch (psi)	pascal (Pa)	6895
Bending Moment or Torque		
inch-pound-force	newton-meter (Nm)	0.1130
foot-pound-force	newton-meter (Nm)	1.356
meter-kilogram-force	newton-meter (Nm)	9.807
Mass		
ounce-mass (avoirdupois)	gram (g)	28.35
pound-mass (avoirdupois)	kilogram (kg)	0.4536
ton (metric)	megagram (Mg)	1.000E
ton (short 2000 lbm)	megagram (Mg)	0.9072
Mass per Volume		
pound-mass/cubic foot	kilogram/cubic meter (kg/m^3)	16.02
pound-mass/cubic yard	kilogram/cubic meter (kg/m^3)	0.5933
pound-mass/gallon	kilogram/cubic meter (kg/m^3)	119.8
Temperature		
deg Fahrenheit (F)	deg Celsius (C)	$t_C = (t_F - 32)/1.8$
deg Celsius (C)	deg Fehrenheit (F)	$t_F = 1.8t_C + 32$

Figure A.1 Selected conversion factors to SI units.

Bar size designa-tion	Nominal cross section area, sq. in.	Weight, lb per ft	Nominal diameter, in.
#3	0.11	0.376	0.375
#4	0.20	0.668	0.500
#5	0.31	1.043	0.625
#6	0.44	1.502	0.750
#7	0.60	2.044	0.875
#8	0.79	2.670	1.000
#9	1.00	3.400	1.128
#10	1.27	4.303	1.270
#11	1.56	5.313	1.410
#14	2.25	7.650	1.693
#18	4.00	13.600	2.257

Figure A.2 Geometrical properties of reinforcing bars.

Areas, A_s (or A'_s) in sq.in.
Columns headed [0][5] contain data for bars of one size in groups of one to ten.
Columns headed [1][2][3][4][5] contain data for bars of two sizes with from one to five of each size.

#4 — size #4 (columns 0, 5) and combination with #3 (columns 1–5)

	0	5		#3: 1	2	3	4	5
1	0.20	1.20		0.31	0.42	0.53	0.64	0.75
2	0.40	1.40		0.51	0.62	0.73	0.84	0.95
3	0.60	1.60		0.71	0.82	0.93	1.04	1.15
4	0.80	1.80		0.91	1.02	1.13	1.24	1.35
5	1.00	2.00		1.11	1.22	1.33	1.44	1.55

#5 — size #5 (columns 0, 5), combination with #4, combination with #3

	0	5	#4: 1	2	3	4	5	#3: 1	2	3	4	5
1	0.31	1.86	0.51	0.71	0.91	1.11	1.31	0.42	0.53	0.64	0.75	0.86
2	0.62	2.17	0.82	1.02	1.22	1.42	1.62	0.73	0.84	0.95	1.06	1.17
3	0.93	2.48	1.13	1.33	1.53	1.73	1.93	1.04	1.15	1.26	1.37	1.48
4	1.24	2.79	1.44	1.64	1.84	2.04	2.24	1.35	1.46	1.57	1.68	1.79
5	1.55	3.10	1.75	1.95	2.15	2.35	2.55	1.66	1.77	1.88	1.99	2.10

#6 — size #6 (columns 0, 5), combination with #5, #4, #3

	0	5	#5: 1	2	3	4	5	#4: 1	2	3	4	5	#3: 1	2	3	4	5
1	0.44	2.64	0.75	1.06	1.37	1.68	1.99	0.64	0.84	1.04	1.24	1.44	0.55	0.66	0.77	0.88	0.99
2	0.88	3.08	1.19	1.50	1.81	2.12	2.43	1.08	1.28	1.48	1.68	1.88	0.99	1.10	1.21	1.32	1.43
3	1.32	3.52	1.63	1.94	2.25	2.56	2.87	1.52	1.72	1.92	2.12	2.32	1.43	1.54	1.65	1.76	1.87
4	1.76	3.96	2.07	2.38	2.69	3.00	3.31	1.96	2.16	2.36	2.56	2.76	1.87	1.98	2.09	2.20	2.31
5	2.20	4.40	2.51	2.82	3.13	3.44	3.75	2.40	2.60	2.80	3.00	3.20	2.31	2.42	2.53	2.64	2.75

#7 — size #7 (columns 0, 5), combination with #6, #5, #4

	0	5	#6: 1	2	3	4	5	#5: 1	2	3	4	5	#4: 1	2	3	4	5
1	0.60	3.60	1.04	1.48	1.92	2.36	2.80	0.91	1.22	1.53	1.84	2.15	0.80	1.00	1.20	1.40	1.60
2	1.20	4.20	1.64	2.08	2.52	2.96	3.40	1.51	1.82	2.13	2.44	2.75	1.40	1.60	1.80	2.00	2.20
3	1.80	4.80	2.24	2.68	3.12	3.56	4.00	2.11	2.42	2.73	3.04	3.35	2.00	2.20	2.40	2.60	2.80
4	2.40	5.40	2.84	3.28	3.72	4.16	4.60	2.71	3.02	3.33	3.64	3.95	2.60	2.80	3.00	3.20	3.40
5	3.00	6.00	3.44	3.88	4.32	4.76	5.20	3.31	3.62	3.93	4.24	4.55	3.20	3.40	3.60	3.80	4.00

#8 — size #8 (columns 0, 5), combination with #7, #6, #5

	0	5	#7: 1	2	3	4	5	#6: 1	2	3	4	5	#5: 1	2	3	4	5
1	0.79	4.74	1.39	1.99	2.59	3.19	3.79	1.23	1.67	2.11	2.55	2.99	1.10	1.41	1.72	2.03	2.34
2	1.58	5.53	2.18	2.78	3.38	3.98	4.58	2.02	2.46	2.90	3.34	3.78	1.89	2.20	2.51	2.82	3.13
3	2.37	6.32	2.97	3.57	4.17	4.77	5.37	2.81	3.25	3.69	4.13	4.57	2.68	2.99	3.30	3.61	3.92
4	3.16	7.11	3.76	4.36	4.96	5.56	6.16	3.60	4.04	4.48	4.92	5.36	3.47	3.78	4.09	4.40	4.71
5	3.95	7.90	4.55	5.15	5.75	6.35	6.95	4.39	4.83	5.27	5.71	6.15	4.26	4.57	4.88	5.19	5.50

#9 — size #9 (columns 0, 5), combination with #8, #7, #6

	0	5	#8: 1	2	3	4	5	#7: 1	2	3	4	5	#6: 1	2	3	4	5
1	1.00	6.00	1.79	2.58	3.37	4.16	4.95	1.60	2.20	2.80	3.40	4.00	1.44	1.88	2.32	2.76	3.20
2	2.00	7.00	2.79	3.58	4.37	5.16	5.95	2.60	3.20	3.80	4.40	5.00	2.44	2.88	3.32	3.76	4.20
3	3.00	8.00	3.79	4.58	5.37	6.16	6.95	3.60	4.20	4.80	5.40	6.00	3.44	3.88	4.32	4.76	5.20
4	4.00	9.00	4.79	5.58	6.37	7.16	7.95	4.60	5.20	5.80	6.40	7.00	4.44	4.88	5.32	5.76	6.20
5	5.00	10.00	5.79	6.58	7.37	8.16	8.95	5.60	6.20	6.80	7.40	8.00	5.44	5.88	6.32	6.76	7.20

#10 — size #10 (columns 0, 5), combination with #9, #8, #7

	0	5	#9: 1	2	3	4	5	#8: 1	2	3	4	5	#7: 1	2	3	4	5
1	1.27	7.62	2.27	3.27	4.27	5.27	6.27	2.06	2.85	3.64	4.43	5.22	1.87	2.47	3.07	3.67	4.27
2	2.54	8.89	3.54	4.54	5.54	6.54	7.54	3.33	4.12	4.91	5.70	6.49	3.14	3.74	4.34	4.94	5.54
3	3.81	10.16	4.81	5.81	6.81	7.81	8.81	4.60	5.39	6.18	6.97	7.76	4.41	5.01	5.61	6.21	6.81
4	5.08	11.43	6.08	7.08	8.08	9.08	10.08	5.87	6.66	7.45	8.24	9.03	5.68	6.28	6.88	7.48	8.08
5	6.35	12.70	7.35	8.35	9.35	10.35	11.35	7.14	7.93	8.72	9.51	10.30	6.95	7.55	8.15	8.75	9.35

#11 — size #11 (columns 0, 5), combination with #10, #9, #8

	0	5	#10: 1	2	3	4	5	#9: 1	2	3	4	5	#8: 1	2	3	4	5
1	1.56	9.36	2.83	4.10	5.37	6.64	7.91	2.56	3.56	4.56	5.56	6.56	2.35	3.14	3.93	4.72	5.51
2	3.12	10.92	4.39	5.66	6.93	8.20	9.47	4.12	5.12	6.12	7.12	8.12	3.91	4.70	5.49	6.28	7.07
3	4.68	12.48	5.95	7.22	8.49	9.76	11.03	5.68	6.68	7.68	8.68	9.68	5.47	6.26	7.05	7.84	8.63
4	6.24	14.04	7.51	8.78	10.05	11.32	12.59	7.24	8.24	9.24	10.24	11.24	7.03	7.82	8.61	9.40	10.19
5	7.80	15.60	9.07	10.34	11.61	12.88	14.15	8.80	9.80	10.80	11.80	12.80	8.59	9.38	10.17	10.96	11.75

#14 — size #14 (columns 0, 5), combination with #11, #10, #9

	0	5	#11: 1	2	3	4	5	#10: 1	2	3	4	5	#9: 1	2	3	4	5
1	2.25	13.50	3.81	5.37	6.93	8.49	10.05	3.52	4.79	6.06	7.33	8.60	3.25	4.25	5.25	6.25	7.25
2	4.50	15.75	6.06	7.62	9.18	10.74	12.30	5.77	7.04	8.31	9.58	10.85	5.50	6.50	7.50	8.50	9.50
3	6.75	18.00	8.31	9.87	11.43	12.99	14.55	8.02	9.29	10.56	11.83	13.10	7.75	8.75	9.75	10.75	11.75
4	9.00	20.25	10.56	12.12	13.68	15.24	16.80	10.27	11.54	12.81	14.08	15.35	10.00	11.00	12.00	13.00	14.00
5	11.25	22.50	12.81	14.37	15.93	17.49	19.05	12.52	13.79	15.06	16.33	17.60	12.25	13.25	14.25	15.25	16.25

#18 — size #18 (columns 0, 5), combination with #14, #11, #10

	0	5	#14: 1	2	3	4	5	#11: 1	2	3	4	5	#10: 1	2	3	4	5
1	4.00	24.00	6.25	8.50	10.75	13.00	15.25	5.56	7.12	8.68	10.24	11.80	5.27	6.54	7.81	9.08	20.35
2	8.00	28.00	10.25	12.50	14.75	17.00	19.25	9.56	11.12	12.68	14.24	15.80	9.27	10.54	11.81	13.08	14.35
3	12.00	32.00	14.25	16.50	18.75	21.00	23.25	13.56	15.12	16.68	18.24	19.80	13.27	14.54	15.81	17.08	18.35
4	16.00	36.00	18.25	20.50	22.75	25.00	27.25	17.56	19.12	20.68	22.24	23.80	17.27	18.54	19.81	21.08	22.35
5	20.00	40.00	22.25	24.50	26.75	29.00	31.25	21.56	23.12	24.68	26.24	27.80	21.27	22.54	23.81	25.08	26.35

Figure A.3   Cross-sectional area of bars for various bar combinations.

Spacing, in.	#3	#4	#5	#6	#7	#8	#9	#10	#11	#14	#18	Spacing, in.
					Cross section area of bar, A_s (or A_s'), in.^{2}							
					Bar size							
4	0.33	0.60	0.93	1.32	1.80	2.37	3.00	3.81	4.68			4
4½	0.29	0.53	0.83	1.17	1.60	2.11	2.67	3.39	4.16	6.00		4½
5	0.26	0.48	0.74	1.06	1.44	1.90	2.40	3.05	3.74	5.40	9.60	5
5½	0.24	0.44	0.68	0.96	1.31	1.72	2.18	2.77	3.40	4.91	8.73	5½
6	0.22	0.40	0.62	0.88	1.20	1.58	2.00	2.54	3.12	4.50	8.00	6
6½	0.20	0.37	0.57	0.81	1.11	1.46	1.85	2.34	2.88	4.15	7.38	6½
7	0.19	0.34	0.53	0.75	1.03	1.35	1.71	2.18	2.67	3.86	6.86	7
7½	0.18	0.32	0.50	0.70	0.96	1.26	1.60	2.03	2.50	3.60	6.40	7½
8	0.17	0.30	0.47	0.66	0.90	1.19	1.50	1.91	2.34	3.38	6.00	8
8½	0.16	0.28	0.44	0.62	0.85	1.12	1.41	1.79	2.20	3.18	5.65	8½
9	0.15	0.27	0.41	0.59	0.80	1.05	1.33	1.69	2.08	3.00	5.33	9
9½	0.14	0.25	0.39	0.56	0.76	1.00	1.26	1.60	1.97	2.84	5.05	9½
10	0.13	0.24	0.37	0.53	0.72	0.95	1.20	1.52	1.87	2.70	4.80	10
10½	0.13	0.23	0.35	0.50	0.69	0.90	1.14	1.45	1.78	2.57	4.57	10½
11	0.12	0.22	0.34	0.48	0.65	0.86	1.09	1.39	1.70	2.45	4.36	11
11½	0.11	0.21	0.32	0.46	0.63	0.82	1.04	1.33	1.63	2.35	4.17	11½
12	0.11	0.20	0.31	0.44	0.60	0.79	1.00	1.27	1.56	2.25	4.00	12
13	0.10	0.18	0.29	0.41	0.55	0.73	0.92	1.17	1.44	2.08	3.69	13
14	0.09	0.17	0.27	0.38	0.51	0.68	0.86	1.09	1.34	1.93	3.43	14
15	0.09	0.16	0.25	0.35	0.48	0.63	0.80	1.02	1.25	1.80	3.20	15
16	0.08	0.15	0.23	0.33	0.45	0.59	0.75	0.95	1.17	1.69	3.00	16
17	0.08	0.14	0.22	0.31	0.42	0.56	0.71	0.90	1.10	1.59	2.82	17
18	0.07	0.13	0.21	0.29	0.40	0.53	0.67	0.85	1.04	1.50	2.67	18

Figure A.4   Area of bars in a 1-foot-wide slab strip.

$$I_g = K_{i4}\left(\frac{1}{12} b_w h^3\right) \qquad\qquad K_{i4} = 1 + (\alpha_b - 1)\beta_h^3 + \frac{3(1 - \beta_h)^2(\beta_h)(\alpha_b - 1)}{1 + \beta_h(\alpha_b - 1)}$$

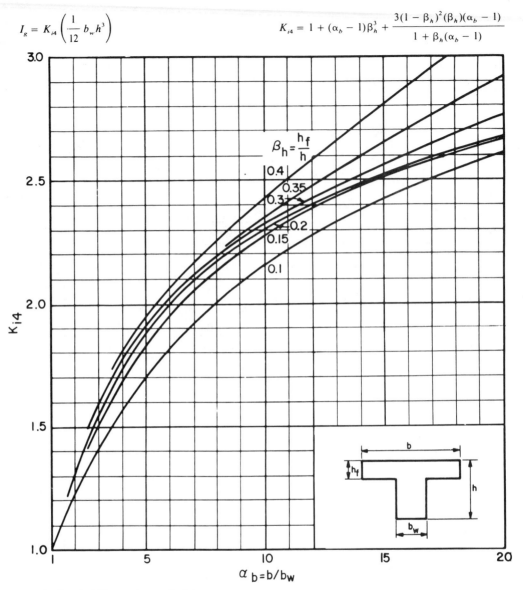

Example: For the T-beam shown, find the moment of inertia I_g:

$$\alpha_b = b/b_w = 143/15 = 9.53$$
$$\beta_h = h_f/h = 8/36 = 0.22$$

Interpolating between the curves for $\beta_h = 0.2$ and 0.3, read $K_{i4} = 2.28$

$$I_g = K_{i4}\frac{b_w h^3}{12} = 2.28\frac{15(36)^3}{12} = 133{,}000 \text{ in.}^4$$

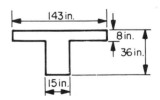

Figure A.5 Gross moment of inertia of T sections.

B

Detailing of Reinforcement

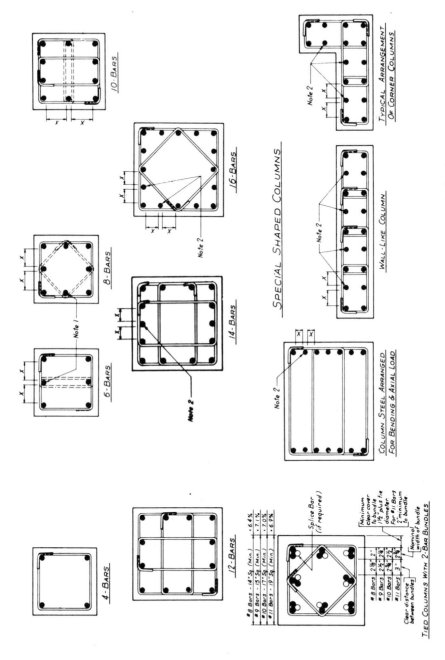

Figure B.1 Column ties for preassembled lap-spliced cages (Ref. 4.2).

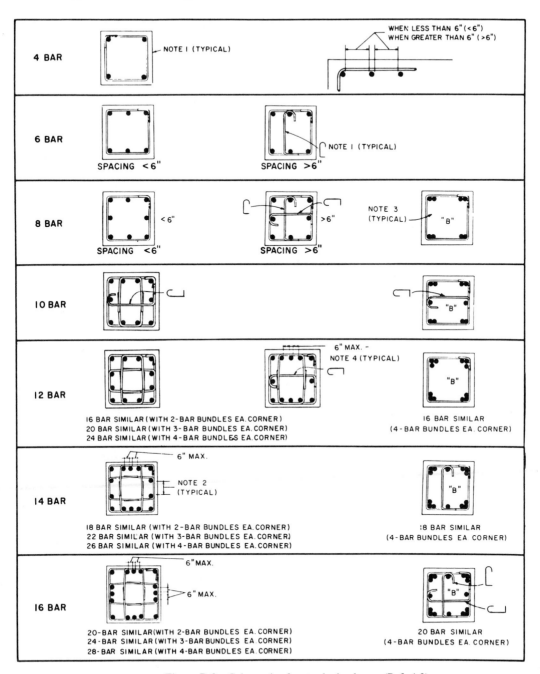

Figure B.2   Column ties for standard columns (Ref. 4.2).

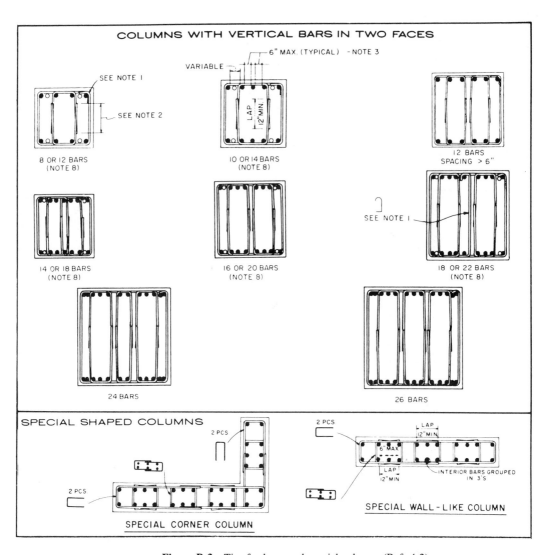

Figure B.3   Ties for large and special columns (Ref. 4.2).

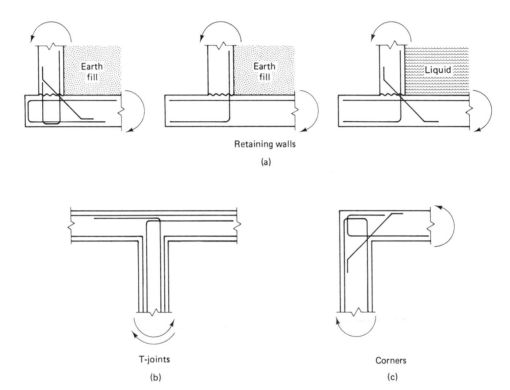

Earth fill

Earth fill

Liquid

Retaining walls

(a)

T-joints

(b)

Corners

(c)

Figure B.4   Corner and joint connection details (Ref. 4.2).

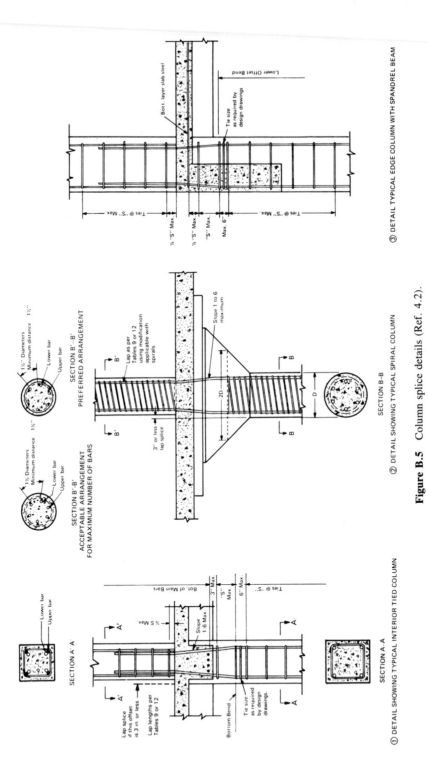

Figure B.5 Column splice details (Ref. 4.2).

① DETAIL SHOWING TYPICAL INTERIOR TIED COLUMN

② DETAIL SHOWING TYPICAL SPIRAL COLUMN

③ DETAIL TYPICAL EDGE COLUMN WITH SPANDREL BEAM

286

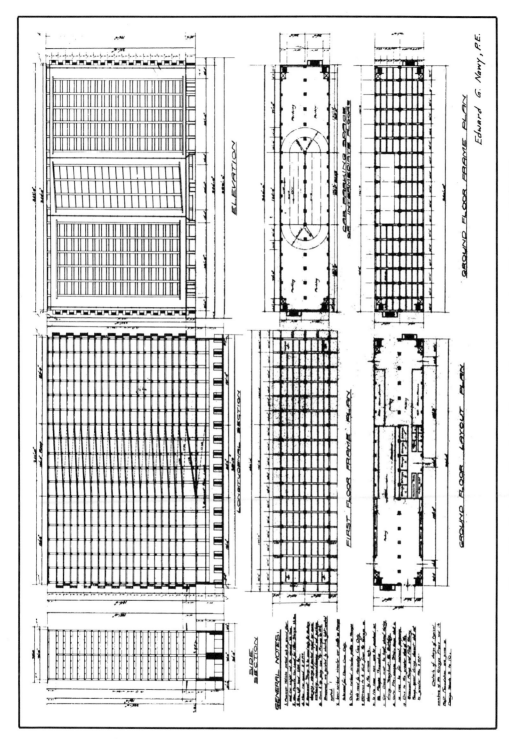

Figure B.6 Typical working drawing for a parking garage structure. (Design by E. G. Nawy.)

287

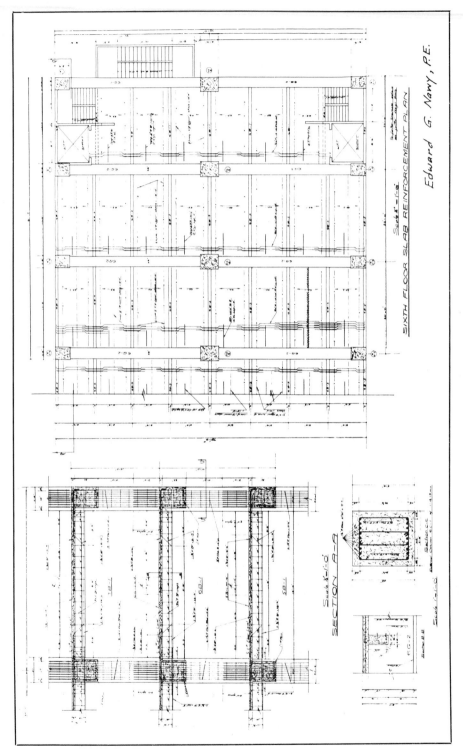

Figure B.7 Typical beam and slab reinforcing working drawing. (Design by E. G. Nawy.)

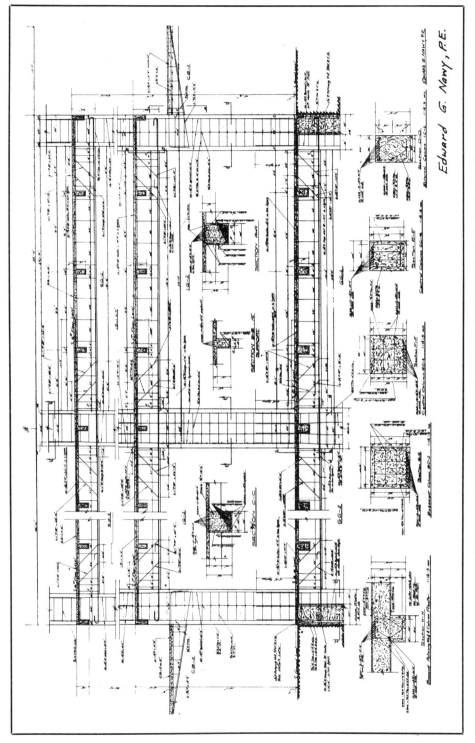

Figure B.8 Typical working drawing of column reinforcing details. (Design by E. G. Nawy.)

289

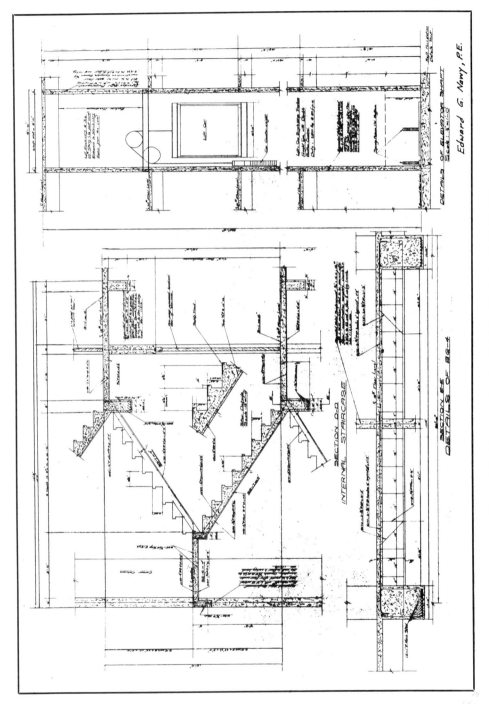

Figure B.9 Elevator and stairwell details. (Design by E. G. Nawy.)

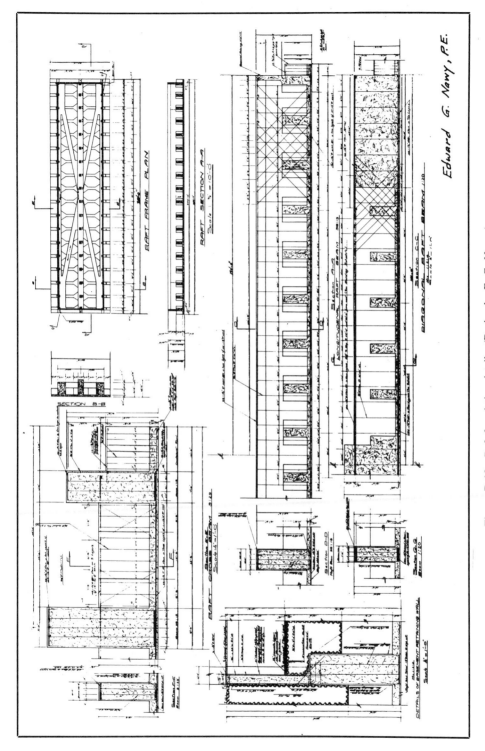

Figure B.10 Raft foundation details. (Design by E. G. Nawy.)

291

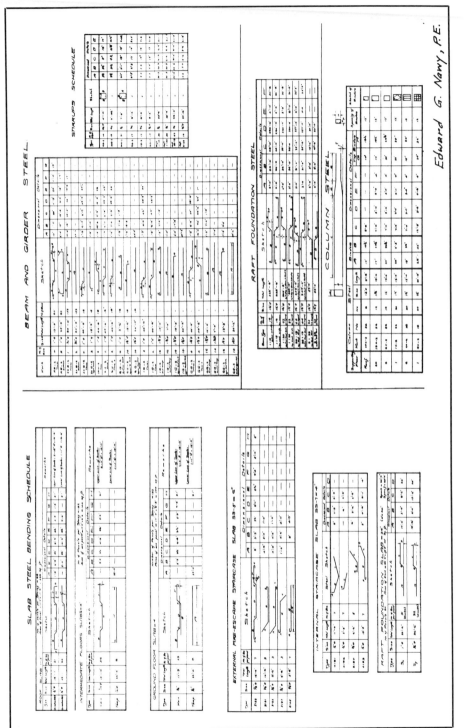

Figure B.11 Typical reinforcement bar bending schedule. (Design by E. G. Nawy.)

APPENDIX

Charts

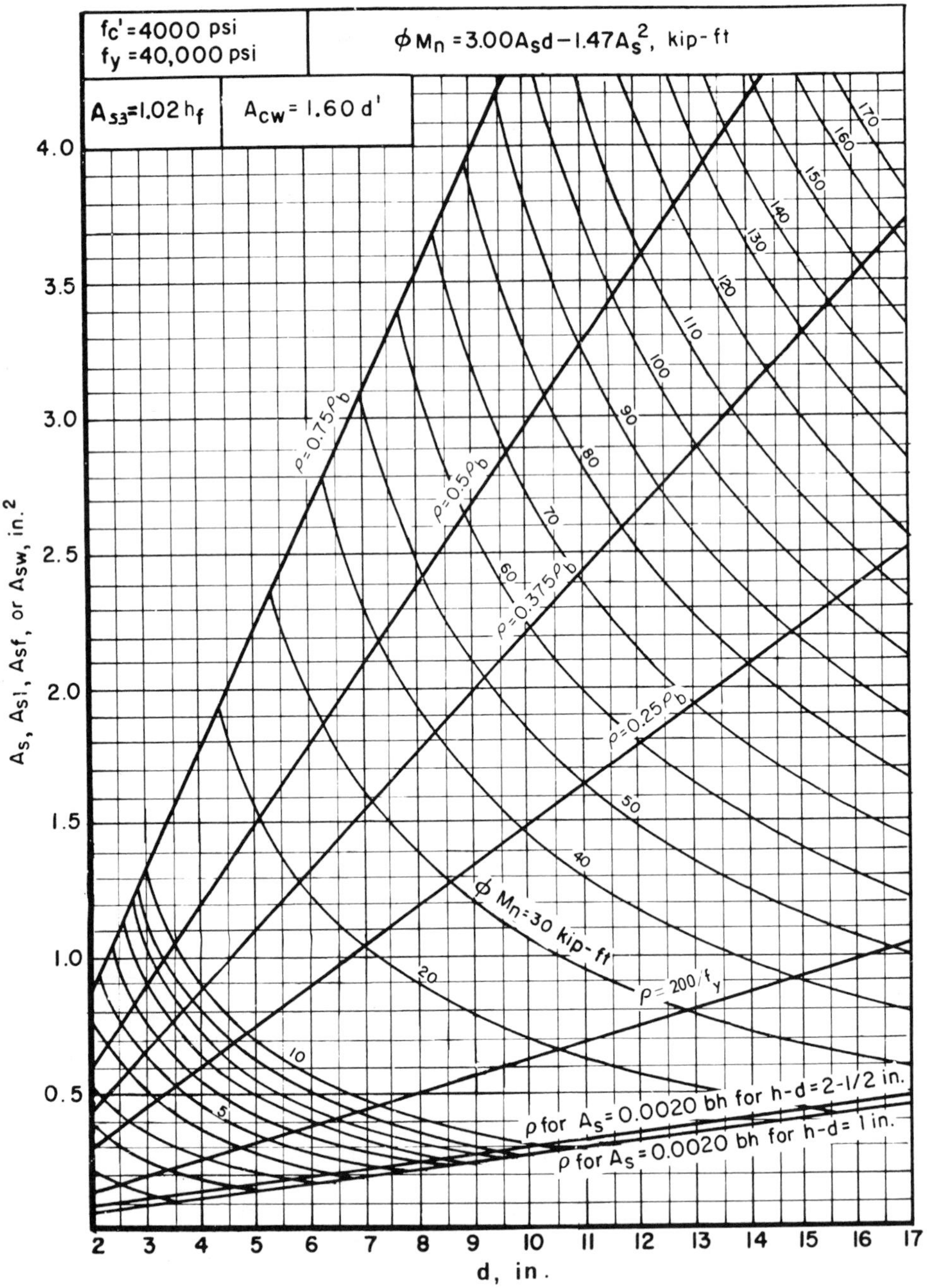

Figure C.1   Design moment strength ϕM_n for slab sections 12 in. wide, $f_c' = 4000$ psi, $f_y = 40,000$ psi (ACI-SP17).

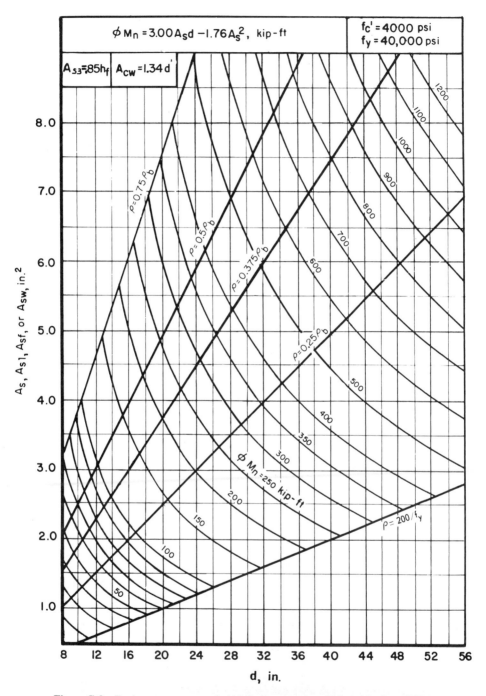

Figure C.2   Design moment strength ϕM_n for beam sections 10 in. wide, $f'_c = 4000$ psi, $f_y = 40,000$ psi (ACI-SP17).

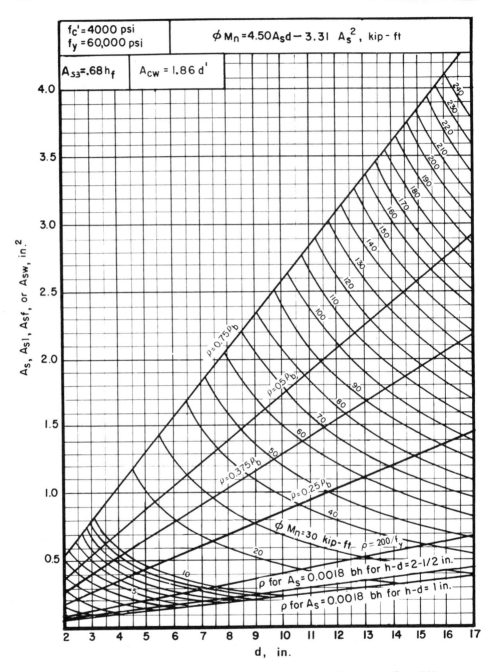

Figure C.3   Design moment strength ϕM_n for slab sections 12 in. wide, $f_c' = 4000$ psi, $f_y = 60,000$ psi (ACI-SP17).

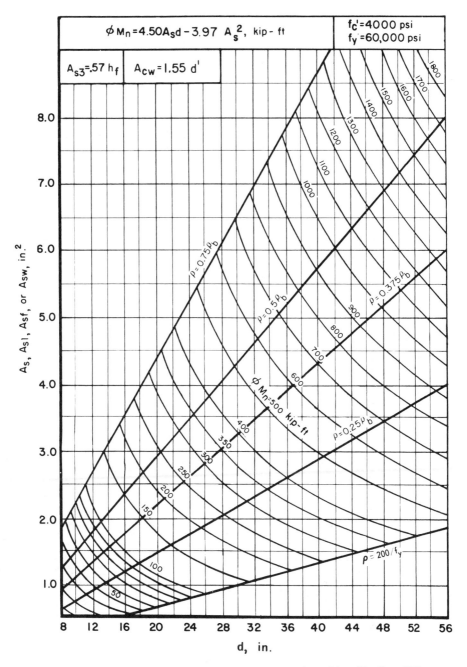

Figure C.4   Design moment strength ϕM_n for beam sections 10 in. wide, $f_c' = 4000$ psi, $f_y = 60,000$ psi (ACI-SP17).

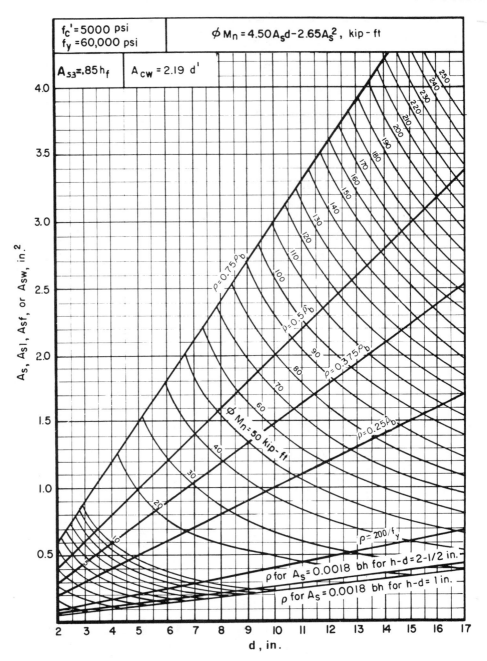

Figure C.5 Design moment strength ϕM_n for slab sections 12 in. wide, $f_c' = 5000$ psi, $f_y = 60,000$ psi (ACI-SP17).

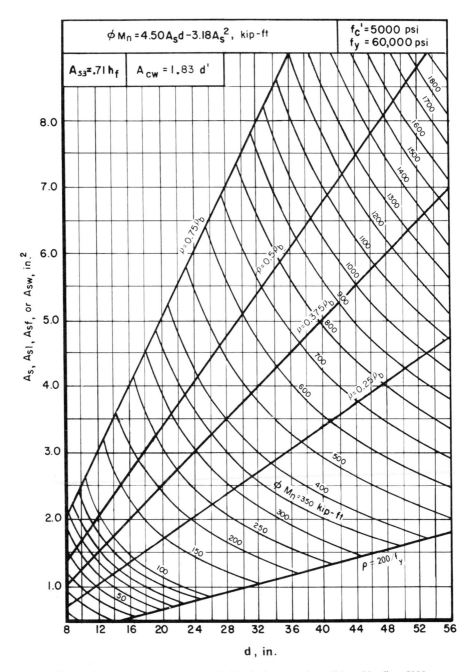

Figure C.6   Design moment strength ϕM_n for beam sections 10 in. wide, $f_c' = 5000$ psi, $f_y = 60,000$ psi (ACI-SP17).

$$\phi V_s = V_u - \phi V_c = \phi A_v f_y \frac{d}{s}$$

$$\text{Maximum } b_w = \frac{A_v f_y}{50s}$$

Stirrup size	s, in. / d, in.	2	3	4	5	6	7	8	9	10	11	12	14	16	18	20
												ϕV_s, kips				
	8	30	20	15												
	10	37	25	19	15											
	12	45	30	22	18	15										
	14	52	35	26	21	17	15									
	16	60	40	30	24	20	17	15								
	18	67	45	34	27	22	19	17	15			Maximum spacing = $d/2$				
	20	75	50	37	30	25	21	19	17	15						
	22	82	55	41	33	27	24	21	18	16	15					
#3 stirrups*	24	90	60	45	36	30	26	22	20	18	16	15				
	26	97	65	49	39	32	28	24	22	19	18	16				
	28	105	70	52	42	35	30	26	23	21	19	17	15			
	30	112	75	56	45	37	32	28	25	22	20	19	16			
	32	120	80	60	48	40	34	30	27	24	22	20	17	15		
	34	127	85	64	51	42	36	32	28	25	23	21	18	16		
	36	135	90	67	54	45	38	34	30	27	24	22	19	17	15	
	38	142	95	71	57	47	41	36	32	28	26	24	20	18	16	
	40	150	100	75	60	50	43	37	33	30	27	25	21	19	17	15
	Max b_w, in.	88	59	44	35	29	25	22	19	17	16	15	13	11	10	9
	8	54	36	27												
	10	68	45	34	27											
	12	82	54	41	33	27										
	14	95	63	48	38	32	27									
	16	109	73	54	44	36	31	27								
	18	122	82	61	49	41	35	31	27			Maximum spacing = $d/2$				
	20	136	91	68	54	45	39	34	30	27						
	22	150	100	75	60	50	43	37	33	30	27					
#4 stirrups*	24	163	109	82	65	54	47	41	36	33	30	27				
	26	177	118	88	71	59	51	44	39	35	32	29				
	28	190	127	95	76	63	54	48	42	38	35	32	27			
	30	204	136	102	82	68	58	51	45	41	37	34	29			
	32	218	145	109	87	73	62	54	48	44	40	36	31	27		
	34	231	154	116	92	77	66	58	51	46	42	38	33	29		
	36	245	163	122	98	82	70	61	54	49	45	41	35	31	27	
	38	258	172	129	103	86	74	65	57	52	47	43	37	32	29	
	40	272	181	136	109	91	78	68	60	54	49	45	39	34	30	27
	Max b_w, in.	160	107	80	64	53	46	40	36	32	29	27	23	20	18	16

Figure C.7   Design shear strength ϕV_s of U stirrups; $f_y = 40{,}000$ psi (ACI-SP17).

$$\phi V_s = V_u - \phi V_c = \phi A_v f_y \frac{d}{s}$$

$$\text{Maximum } b_w = \frac{A_v f_y}{50s}$$

Stirrup size	s, in. / d, in.	ϕV_s, kips														
		2	3	4	5	6	7	8	9	10	11	12	14	16	18	20
	8	45	30	22												
	10	56	37	28	22											
	12	67	45	34	27	22										
	14	79	52	39	31	26	22									
	16	90	60	45	36	30	26	22								
	18	101	67	50	40	34	29	25	22							
	20	112	75	56	45	37	32	28	25	22						
	22	123	82	62	49	41	35	31	27	25	22					
#3 stirrups*	24	135	90	67	54	45	38	34	30	27	24	22				
	26	146	97	73	58	49	42	36	32	29	27	24				
	28	157	105	79	63	52	45	39	35	31	29	26	22			
	30	168	112	84	67	56	48	42	37	34	31	28	24			
	32	180	120	90	72	60	51	45	40	36	33	30	26	22		
	34	191	127	95	76	64	54	48	42	38	35	32	27	24		
	36	202	135	101	81	67	58	50	45	40	37	34	29	25	22	
	38	213	142	107	85	71	61	53	47	43	39	36	30	27	24	
	40	224	150	112	90	75	64	56	50	45	41	37	32	28	25	22
	Max b_w, in.	32	88	66	52	44	37	33	29	26	24	22	18	16	15	13
	8	82	54	41												
	10	102	68	51	41											
	12	122	82	61	49	41										
	14	143	95	71	57	48	41									
	16	163	109	82	65	54	47	41								
	18	184	122	92	73	61	52	46	41							
	20	204	136	102	82	68	58	51	45	41						
	22	224	150	112	90	75	64	56	50	45	41					
#4 stirrups*	24	245	163	122	98	82	70	61	54	49	45	41				
	26	265	177	133	106	88	76	66	59	53	48	44				
	28	286	190	143	114	95	82	71	63	57	52	48	41			
	30	306	204	153	122	102	87	77	68	61	56	51	44			
	32	326	207	163	131	109	93	82	73	65	59	54	47	41		
	34	347	231	173	139	116	99	87	77	69	63	58	50	43		
	36	367	245	184	147	122	105	92	82	73	67	61	52	46	41	
	38	388	258	194	155	129	111	97	86	78	70	65	55	48	43	
	40	408	272	204	163	136	117	102	91	82	74	68	58	51	45	41
	Max b_w, in.	240	160	120	96	80	68	60	53	48	44	40	34	30	27	24

(Maximum spacing = $d/2$)

Figure C.8   Design shear strength ϕV_s of U stirrups; $f_y = 60,000$ psi (ACI-SP17).

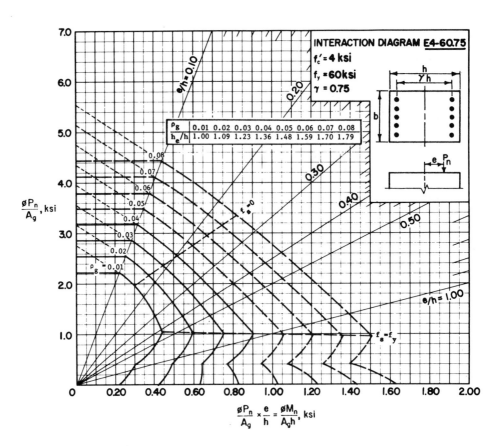

Figure C.9 Rectangular columns load-moment strength interaction diagrams; $f_c' = 4000$ psi, $f_y = 60,000$ psi, $\gamma = 0.75$ (ACI-SP17).

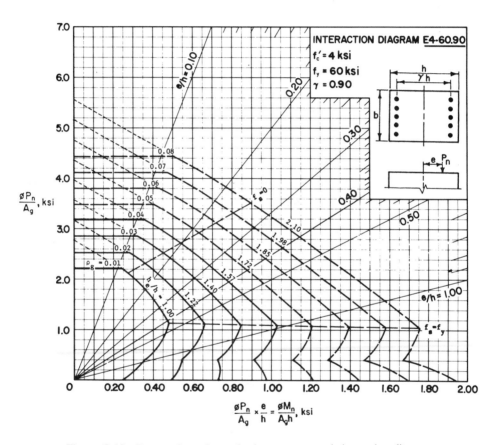

Figure C.10 Rectangular columns load-moment strength interaction diagrams, $f_c' = 4000$ psi, $f_y = 60,000$ psi, $\gamma = 0.90$ (ACI-SP17).

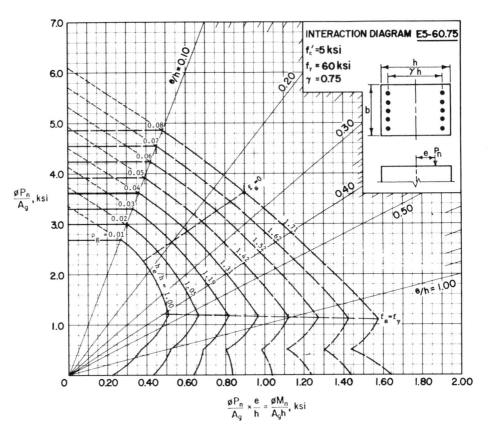

Figure C.11 Rectangular columns load-moment strength interaction diagrams, $f'_c = 5000$ psi, $f_y = 60,000$ psi, $\gamma = 0.75$ (ACI-SP17).

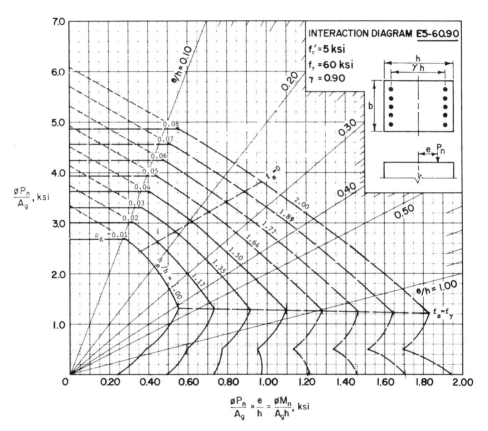

Figure C.12 Rectangular columns load-moment strength interaction diagrams, $f_c' = 5000$ psi, $f_y = 60,000$ psi, $\gamma = 0.90$ (ACI-SP17).

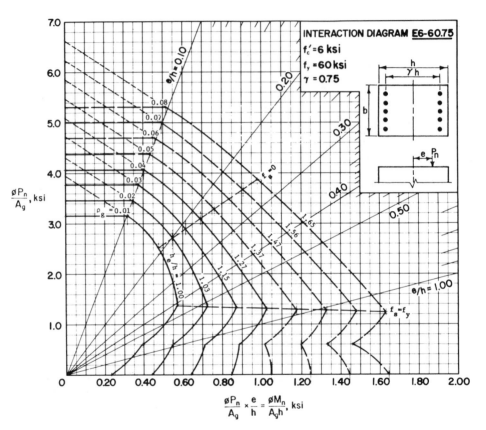

Figure C.13 Rectangular columns load-moment strength interaction diagrams, $f'_c = 6000$ psi, $f_y = 60,000$ psi, $\gamma = 0.75$ (ACI-SP17).

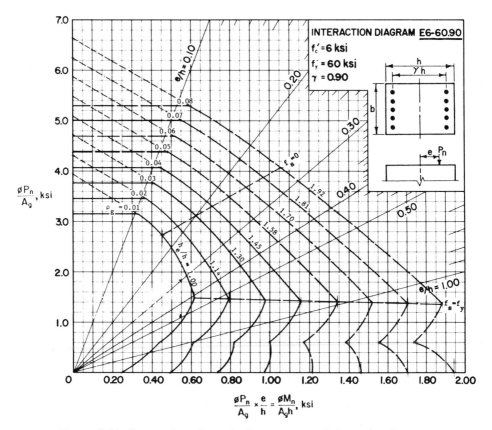

Figure C.14 Rectangular columns load-moment strength interaction diagrams, $f'_c = 6000$ psi, $f_y = 60,000$ psi, $\gamma = 0.90$ (ACI-SP17).

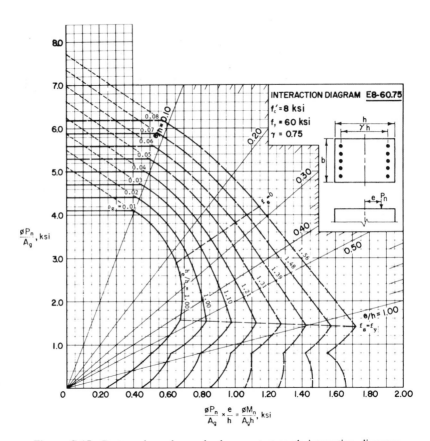

Figure C.15 Rectangular columns load-moment strength interaction diagrams, $f_c' = 8000$ psi, $f_y = 60,000$ psi, $\gamma = 0.75$ (ACI-SP17).

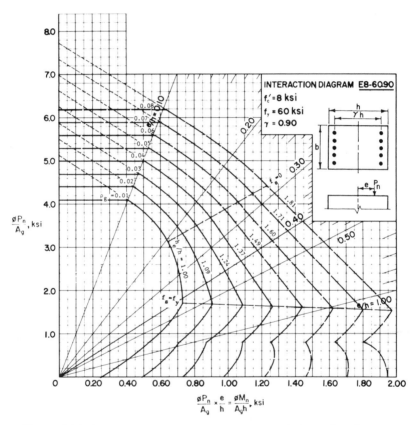

Figure C.16 Rectangular columns load-moment strength interaction diagrams, $f'_c = 8000$ psi, $f_y = 60,000$ psi, $\gamma = 0.90$ (ACI-SP17).

Index

A

Accelerating admixtures, 17
Adhesion (*see* Bond development)
Admixtures, 17
Aggregates, 13, 14
Air entrainment, 15
American Concrete Institute, 21, 42, 49, 96, 112, 116, 153, 239, 283–86, 375, 418, 497, 498
American Insurance Association, 273
American Society for Testing and Materials, 42
Anchorage bond, 101
Axial load combined with bending, 198, 204
Axially loaded columns, 200, 201

B

Balaguru, P. N., 42
Balanced reinforcement ratio, 55
Balanced section, 55
Bar sizes, 36, 37
Beams:
  balanced section, 55
  beams without diagonal tension reinforcement, 123
  bond, 101, 102
  compression failure, 207
  deflections, 157, 166, 170, 178
  diagonal tension analysis, 118
  doubly reinforced, 70
  flexural cracking, 181
  homogeneous beams, 118
  maximum reinforcement percentages, 55
  minimum reinforcement percentage, 55, 73
  nominal moment of strength, 51, 85, 107
  nonhomogeneous section, 123
  nonrectangular sections, 107
  overreinforced section, 53
  shear strength, 123, 126
  singly reinforced rectangular beams:
    analysis, 56
    design, 62
  strain-distribution, 46, 50
  T and L beams, 80, 81, 83, 86, 87
  torsion, 136
Bearing capacity of soil, 242

Bending:
  combined with axial load, 204
  combined with torsion and shear, 140
  (*see also* Beams; One-way slabs)
Blair, K. W., 192
Bond stress, 102
Branson, D. E., 191
Bundled bars, 183

C

Columns:
  balanced failure, 207
  circular columns, 203, 232
  compression failure, 207, 213, 215
  concentrically loaded columns, 200
  eccentrically loaded columns, 204
  lateral reinforcement, 197, 231, 232
  load moment strength column interaction diagram, 223
  nominal strength reduction factor, 48, 221
  types of columns, 196
  Whitney's approximate solution, 219
Compressive stress block, 46
Concrete:
  admixtures, 17
  aggregates, 15
  air-content, 14
  compressive strength, 21, 27
  creep, 34
  curing, 26
  entrained air, 17
  historical development, 1
  lightweight concrete, 14, 32
  modulus of elasticity, 31
  Portland cement, 13
  shrinkage, 33
  slump, 26
  stress-strain curve, 29
  tensile strength, 24, 27, 29
  water, 14
Concrete structural systems:
  beams, 39
  columns, 40
  foundations, 41
Cover for steel reinforcement, 36

Crack control:
  beams and one-way slabs, 181
  two-way slabs and plates, 186
Cracking moment, 160
Creep, 34

D

Deflections:
  deflection behavior of beams, 158
  deflection requirements for minimum thickness, 169, 170
  long-term deflection, 166
  maximum deflection expressions, 173, 174
  postcracking stage, 159
Development length, 100, 102, 104
Diagonal tension, 118, 123, 124
Doubly reinforced beam sections, 70

E

Eccentrically loaded columns, 204
Effective moment of inertia, 165
Effective width of T section, 83
Elasticity modulus, 31
Embedment length, 106, 107, 108
Equilibrium torsion, 140
Equivalent rectangular block, 49

F

Ferguson, P. M., 116
Flanged beams, 80, 81
Flexural cracking, 181
Flexural members (*see* Beams; One-way slabs)
Flow-charts:
  analysis of doubly reinforced rectangular beam, 75
  analysis of singly reinforced rectangular beam in bending, 57
  analysis of T and L beams, 86
  deflection evaluation, 172
Footings:
  bearing capacity, 241
  combined footing, 265
  design considerations in flexure, 251
  design considerations in shear:
    beam action, 253
    two-way action, 253
  eccentrically loaded footings, 249

Footings (*cont.*):
  operation procedure for the design of footings, 255
  reinforcement distribution, 251
  shear behavior, 245
Fracture coefficient, 188
Furlong, R. W., 273

H

Hsu, T. T. C., 153

L

Lambot, J. L., 2
Limit state of failure in flexure, 82
Load factors, 70
Loads, 69, 70, 82

M

Maximum allowable concrete strain, 50
Maximum reinforcement percentage:
  beams, 54
  columns, 231
Members in flexure (*see* Beams)
Members in flexure and axial load (*see* Columns)
Mindness, S., 55
Minimum reinforcement percentage:
  beams, 54
  columns, 231
Modulus of elasticity:
  concrete, 31
  steel, 35
Modulus of rupture of concrete, 29
Mohr's circle, 121
Moment of inertia, 159, 160, 161
Monier, J., 2

N

Nawy, E. G., 42, 96, 191, 192
Nominal moment strength, 51
Nominal shear strength, 152, 157
Nominal strength, 71
Nominal torsional strength, 204
Nonrectangular sections, 80, 81

O

One-way slabs, 68

P

Portland cement, 13
Portland Cement Association, 13

R

Reinforcement:
  anchorage (*see* Bond stress)
  area, cross section, 36, 37
  bond development, 100
  cover, 39
  flexural reinforcement (*see* Beams; One-way slabs)
  grade, 36, 37
  ratio, 52
  stirrups, size and spacing, 130
  web reinforcement, 130
  weights of deformed bars, 37
  wires, 38
  yield strength, 37
  Young's modulus, 37

S

Safety factors, 41, 42
Serviceability, 146
Shear strength:
  behavior of homogeneous beams, 118
  behavior of nonhomogeneous beams, 123
  nominal shear strength, 124, 125
  shear strength of plain concrete, 125
  spacing of stirrups, 130
  web reinforcement design, 131
Shrinkage, 33
Slabs (*see* One-way slabs)
Slurry, 9
Soil pressure, 242
Splices, 111
Steel (*see* Reinforcement)
Stirrups (*see* Web-reinforcement)
Strain-compatibility check, 72
Strength of concrete:
  compressive, 29, 30
  tensile, 27

Strength reduction factor, 49
Stress-strain curve:
  concrete, 30, 198
  steel, 36

T

T beams, 80, 81
Tests on concrete, quality, 17, 21
Ties, lateral, 197, 231, 232
Torsion:
  combined torsion, bending, shear, 140
  compatibility torsion, 141
  equilibrium torsion, 140
  pure torsion, 137, 138
  web-reinforcement, 143
Trajectories of stresses, 123
Truss analogy, planar, 126

W

Wall footings, 243
Water, 14
Water/cement ratio, 22, 25
Water-reducing admixtures, 17
Web-reinforcement:
  shear, 129, 130
  torsion, 147
Whitney, C. S., 96, 239

Y

Yield strength, 37